Study Guide

to accompany

Pytel • Kiusalaas

ENGINEERING MECHANICS
Statics

Study Guide

to accompany

Pytel • Kiusalaas

ENGINEERING MECHANICS
Statics

Prepared by

Jean Landa Pytel
The Pennsylvania State University

HarperCollins*CollegePublishers*

Study Guide to accompany Pytel/Kiusalaas, ENGINEERING MECHANICS: STATICS

ISBN 0-06-501946-6

94 95 96 97 9 8 7 6 5 4 3 2 1

PREFACE

To the Student

The goal of this Study Guide written to accompany *STATICS,* by Pytel and Kiusalaas, is to help you master the principles of statics.

There is no discussion in the Study Guide of material that is labeled supplementary in the text. This Study Guide is not designed to replace or add to the text but to help you focus on the key concepts and skills presented in the text. To accomplish this goal, each section of the Study Guide is divided into three parts.

> A. *You Should Understand* highlights the key concepts covered in the section.
>
> B. *You Should be Able to* will help you identify the skills necessary to solve the homework problems.
>
> C. *Guided Practice Problems* includes one or more guided practice problems that reinforce problem solving skills. The guided practice problems will give you an opportunity to work through at least one problem with some guidance, before attempting the homework problems. The icon ✍ indicates where you are to perform calculations and fill in the blanks. The answers are provided in a box at the end of each guided problem. The solution to each guided problem can be found in the Solution Section.

CONTENTS

S 1. FUNDAMENTAL PROPERTIES OF VECTORS

Text Reference: Article 1.3; Sample Problems 1.1, 1.2

A. You Should Understand:

- Two vectors **A** and **B** are equal (**A** = **B**) if their magnitudes are equal, $|\mathbf{A}| = |\mathbf{B}|$, and they have the same direction.

- If **A** = –**B**, the two vectors have the same magnitude, $|\mathbf{A}| = |\mathbf{B}|$, but they have opposite directions.

- A <u>unit vector</u> is a dimensionless vector with a magnitude of 1, frequently represented by $\boldsymbol{\lambda}$.

  Note: (i) $|\boldsymbol{\lambda}| = 1$.

  (ii) If $\boldsymbol{\lambda}$ is in the same direction as **A**, then $\mathbf{A} = A\boldsymbol{\lambda}$, where $A = |\mathbf{A}|$.

- Addition of vectors obeys the Parallelogram Law of Addition. If **C** = **A** + **B**, **C** is called the resultant of **A** and **B**, whereas **A** and **B** are the components of **C**.

- The subtraction of two vectors: $\mathbf{A} - \mathbf{B} = \mathbf{A} + (-\mathbf{B})$.

B. You Should be Able to:

- Use the Parallelogram Law of Addition to add any two vectors.

C. Guided Practice Problem

S 1.1 Determine analytically, the resultant **R** of the two forces shown. Use P = 150 N and Q = 75 N.

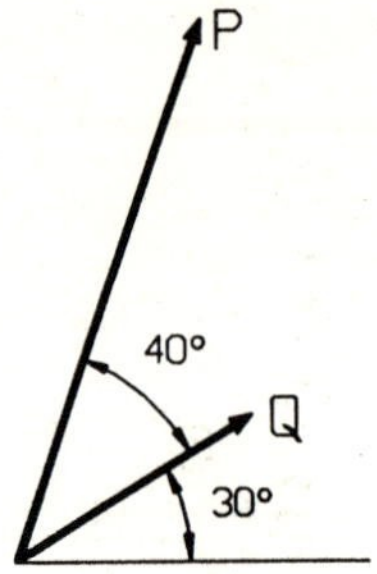

GUIDED SOLUTION

To determine the magnitude and direction of **R** analytically, we first sketch the force triangle, not necessarily to scale.

1. Determine the angle β. Below, we have drawn the force triangle by putting the tail of **P** at the head of **Q** and closing the triangle with **R**.

✍

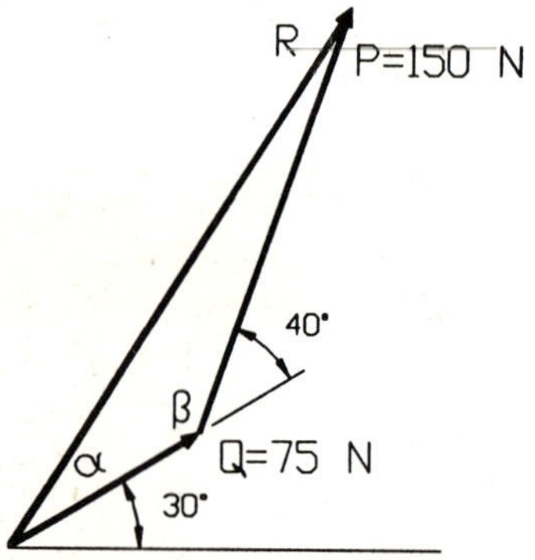

$\beta =$

2. Calculate the magnitude of **R**. Use the Law of Cosines.

✍ $R = \sqrt{P^2 + Q^2 - 2\,P\,Q\cos\beta} =$

3. Determine the angle $(\alpha + 30°)$ that **R** makes with the horizontal.

✍ (a) Using the Law of Sines, solve for α:

$$\frac{R}{\sin\beta} = \frac{P}{\sin\alpha}$$

$$\sin\alpha = \frac{P\sin\beta}{R} =$$

$$\therefore \alpha =$$

✍ (b) Calculate: $\alpha + 30° =$

| **R** | = 213 N at 56.9° from the horizontal

S 2. REPRESENTATION OF VECTORS USING RECTANGULAR COMPONENTS

Text Reference: Article 1.4; Sample Problems 1.3-1.5

A. You Should Understand:

- Vector **A** can be written as the sum of its three rectangular components: $\mathbf{A} = A_x\mathbf{i} + A_y\mathbf{j} + A_z\mathbf{k}$.

- Using rectangular components, the addition of vectors becomes: $\mathbf{A} + \mathbf{B} = (A_x + B_x)\mathbf{i} + (A_y + B_y)\mathbf{j} + (A_z + B_z)\mathbf{k}$.

- The <u>direction cosines</u> of **A** are the cosines of the angle between **A** and each of the <u>positive coordinate axes</u>: $\cos\theta_x = \frac{A_x}{A}$, $\cos\theta_y = \frac{A_y}{A}$, $\cos\theta_z = \frac{A_z}{A}$.

- Vector **A** may be written in terms of its direction cosines: $\mathbf{A} = A\cos\theta_x\,\mathbf{i} + A\cos\theta_y\,\mathbf{j} + A\cos\theta_z\,\mathbf{k}$. The terms $A_x = A\cos\theta_x$, $A_y = A\cos\theta_y$, and $A_z = A\cos\theta_z$ are called the scalar components of **A**.

- A unit vector $\boldsymbol{\lambda}$ can be written as: $\boldsymbol{\lambda} = \cos\theta_x\mathbf{i} + \cos\theta_y\mathbf{j} + \cos\theta_z\mathbf{k}$. The direction cosines of the unit vector are $\cos\theta_x$, $\cos\theta_y$, and $\cos\theta_z$.

B. You Should be Able to:

- Write a vector between any two points: $\overrightarrow{AB} = (x_B - x_A)\mathbf{i} + (y_B - y_A)\mathbf{j} + (z_B - z_A)\mathbf{k}$, where $\overrightarrow{AB}$ is a vector from point $A(x_A, y_A, z_A)$ to point $B(x_B, y_B, z_B)$.

- Write a vector in terms of its scalar components: $\mathbf{A} = A_x\mathbf{i} + A_y\mathbf{j} + A_z\mathbf{k}$.

- Calculate the magnitude of a vector: $A = |\mathbf{A}| = \sqrt{A_x^2 + A_y^2 + A_z^2}$.

- Add vectors using rectangular components: $\mathbf{A} + \mathbf{B} = (A_x + B_x)\mathbf{i} + (A_y + B_y)\mathbf{j} + (A_z + B_z)\mathbf{k}$.

- Write a unit vector in the direction from A toward B: $\lambda_{AB} = \dfrac{\overrightarrow{AB}}{|\overrightarrow{AB}|}$.

Note: The steps involved in writing the unit vector, λ_{AB}, from point A toward point B are as follows:

1. Write a vector, $\overrightarrow{AB}$, from point A to point B.

2. Calculate the magnitude, $|\overrightarrow{AB}|$, of that vector.

3. Determine the unit vector: $\lambda_{AB} = \dfrac{\overrightarrow{AB}}{|\overrightarrow{AB}|}$.

- Write a vector as the product of its magnitude and a unit vector: $\mathbf{A} = A\lambda_A$.
- Determine the direction cosines of vector **A**: $\cos\theta_x = \dfrac{A_x}{A}$, $\cos\theta_y = \dfrac{A_y}{A}$, and $\cos\theta_z = \dfrac{A_z}{A}$.

C. Guided Practice Problems

S 2.1 The magnitude of the force **F** acting in the direction shown is 500 N.

(a) Write **F** in vector form.

(b) Determine the direction cosines of **F**.

(c) Find the angle that **F** makes with each of the positive coordinate axes.

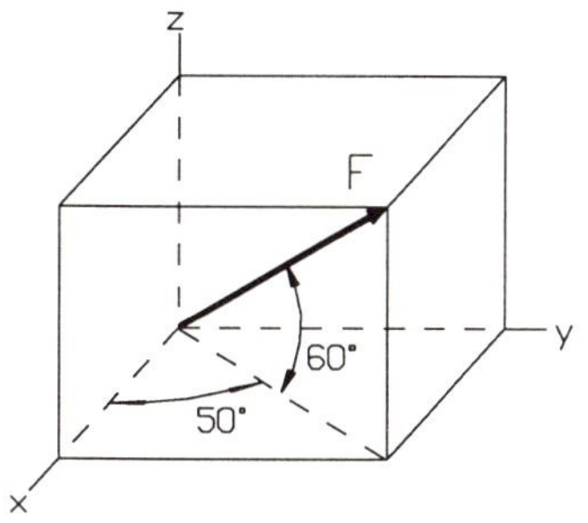

GUIDED SOLUTION

Part (a)

Since the direction of **F** is given, we will determine the scalar components of **F**, and then use them to write **F** in vector form. Note that it is convenient to find the component of **F** in the x-y plane before determining the x and y components of **F**.

1. Determine the components of **F** in the z-direction and in the x-y plane, shown below.

✍

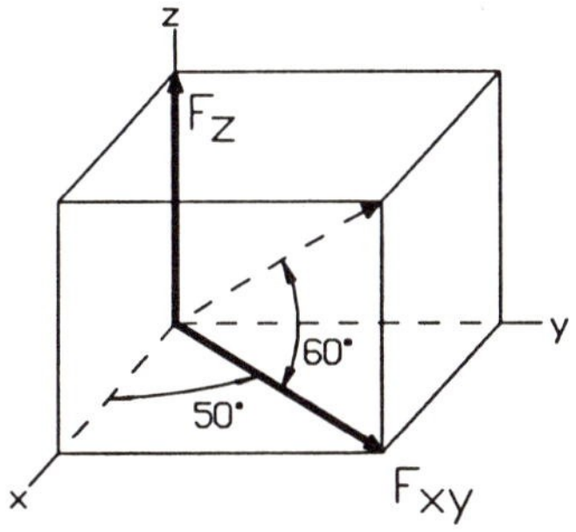

$F_z = F\sin 60° =$

$F_{xy} = F\cos 60° =$

2. Determine the components of **F** in the x- and in the y-directions, shown below.

✍

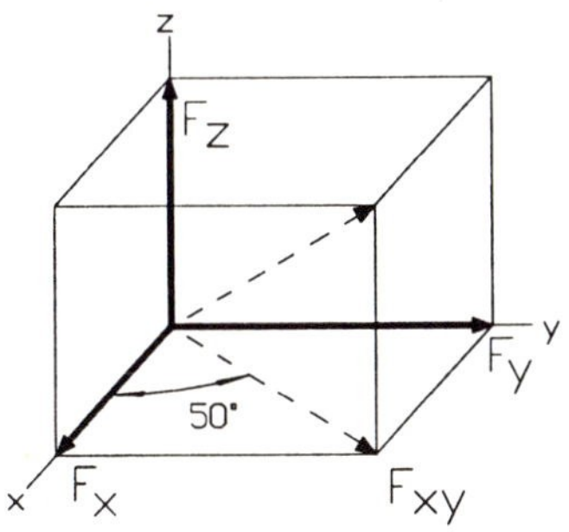

$F_x = F_{xy} \cos 50° =$

$F_y = F_{xy} \sin 50° =$

3. Write **F** in vector form. Using the scalar components determined above, the force **F** can be written as:

✍ $\mathbf{F} = F_x\mathbf{i} + F_y\mathbf{j} + F_z\mathbf{k} =$

F = 161 **i** + 192 **j** + 433 **k** N

Part (b)

1. Determine the direction cosines of **F**. The scalar components of **F** are used for the direction cosines.

✍ $\cos\theta_x = \dfrac{F_x}{F} =$ $\cos\theta_y = \dfrac{F_y}{F} =$

$\cos\theta_z = \dfrac{F_z}{F} =$

$\cos\theta_x = 0.3214$, $\cos\theta_y = 0.3830$, $\cos\theta_z = 0.8660$

Part (c)

1. Find the angle that **F** makes with each of the positive coordinate axes.

✍ $\theta_x = \cos^{-1}\dfrac{F_x}{F} =$ $\theta_y = \cos^{-1}\dfrac{F_y}{F} =$

$\theta_z = \cos^{-1}\dfrac{F_z}{F} =$

$\theta_x = 71.3°$, $\theta_y = 67.5°$, $\theta_z = 30.0°$

S 2.2 Force **P** acts at A and is directed toward B; force **Q** acts at A and is directed toward C. The magnitude of **P** is 130 lb and the magnitude of **Q** is 50 lb.
(a) Write the unit vector directed from A toward B.
(b) Write **P** in vector form.
(c) Write **Q** in vector form.
(d) Determine **R**, the resultant of **P** and **Q**.

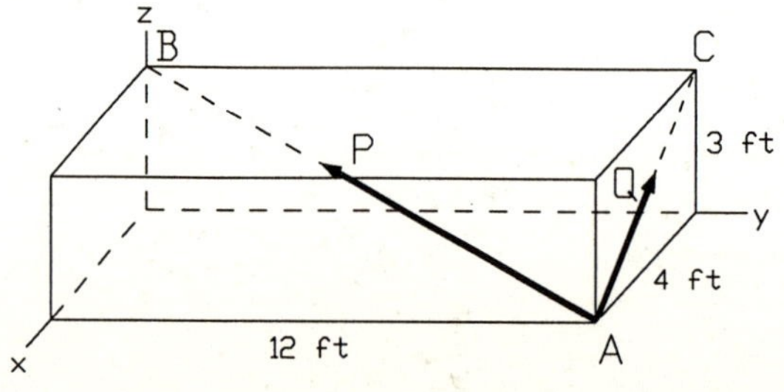

GUIDED SOLUTION

Part (a)

1. Write $\overrightarrow{AB}$, the vector from point A to point B.

✍ $\overrightarrow{AB} =$

2. Calculate the magnitude of $\overrightarrow{AB}$.

✍ $|\overrightarrow{AB}| =$

3. Determine the unit vector λ_{AB}.

✍ $\lambda_{AB} = \dfrac{\overrightarrow{AB}}{|\overrightarrow{AB}|} =$

$\boxed{\lambda_{AB} = -0.3077\,\mathbf{i} - 0.9231\,\mathbf{j} + 0.2308\,\mathbf{k}}$

Part (b)

1. Write **P** in vector form. Since **P** is in the same direction as $\overrightarrow{AB}$, λ_{AB} can be used to write **P**.

✍ $\mathbf{P} = P\lambda_{AB} =$

$\boxed{\mathbf{P} = -40.0\,\mathbf{i} - 120\,\mathbf{j} + 30.0\,\mathbf{k}\ \text{lb}}$

Part (c)

To write **Q** in vector form, $Q\lambda_{AC}$ is needed. Therefore, λ_{AC} must be determined first.

1. Write $\overrightarrow{AC}$, the vector from point A to point C.

✍ $\overrightarrow{AC} =$

2. Calculate the magnitude of $\overrightarrow{AC}$.

✍ $|\overrightarrow{AC}| =$

3. Determine the unit vector λ_{AC}.

✍ $\lambda_{AC} = \dfrac{\overrightarrow{AC}}{|\overrightarrow{AC}|} =$

4. Write **Q** in vector form.

✍ $\mathbf{Q} = Q\lambda_{AC} =$

$\boxed{\mathbf{Q} = -40.0\,\mathbf{i} + 30.0\,\mathbf{k}\ \text{lb}}$

Part (d)

1. Add **P** and **Q**.

✍ $\mathbf{R} = \mathbf{P} + \mathbf{Q} = (P_x + Q_x)\mathbf{i} + (P_y + Q_y)\mathbf{j} + (P_z + Q_z)\mathbf{k} =$

2. Determine the magnitude of **R**.

✍ $|\mathbf{R}| = \sqrt{R_x^2 + R_y^2 + R_z^2} =$

$\mathbf{R} = -80.0\,\mathbf{i} - 120\,\mathbf{j} + 60.0\,\mathbf{k}$ lb
$\lvert\mathbf{R}\rvert = 156$ lb

S 3. VECTOR MULTIPLICATION

Text Reference: Article 1.5; Sample Problem 1.6

A. You Should Understand:

- The dot product of two vectors **A** and **B** is a scalar given by: $\mathbf{A} \bullet \mathbf{B} = A\,B\cos\theta$, where θ is the angle between **A** and **B** $(0 \le \theta \le 2\pi$ rad).

  Note: The orthogonal component of **B** in the direction of **A** is $B\cos\theta$.

- The cross product of the vectors **A** and **B** is a vector: $\mathbf{A} \times \mathbf{B} = \mathbf{C}$.

  Note: (i) The magnitude of **C** is: $C = A\,B\sin\theta$, where θ is the angle between the positive directions of **A** and **B** $(0 \le \theta \le \pi$ rad).

  (ii) **C** is perpendicular to both **A** and **B**.

  (iii) The direction of **C** is determined by the right hand rule.

- The scalar triple product of the three vectors **A**, **B**, and **C**, is $\mathbf{A} \times \mathbf{B} \bullet \mathbf{C}$.

- The dot product is a scalar, the cross product is a vector, and the scalar triple product is a scalar.

B. You Should be Able to:

- Calculate the dot product of two vectors.

- Find the angle between two vectors. For the angle θ between **A** and **B**:
  $\cos\theta = \dfrac{\mathbf{A} \bullet \mathbf{B}}{AB} = \dfrac{\mathbf{A}}{A} \bullet \dfrac{\mathbf{B}}{B} = \lambda_A \bullet \lambda_B$, where λ_A and λ_B are the unit vectors in the direction of **A** and **B**, respectively.

- Use the dot product to determine the component of one vector in the direction of another vector. The component of **B** in the direction of **A** is: $B\cos\theta = \mathbf{B} \bullet \lambda_A$.

- Determine the dot product: $\mathbf{A} \bullet \mathbf{B} = A_x B_x + A_y B_y + A_z B_z$.

- Determine the cross product of two vectors: $\mathbf{A} \times \mathbf{B} = \begin{vmatrix} \mathbf{i} & \mathbf{j} & \mathbf{k} \\ A_x & A_y & A_z \\ B_x & B_y & B_z \end{vmatrix}$.

- Calculate the scalar triple product: $\mathbf{A} \times \mathbf{B} \bullet \mathbf{C} = \begin{vmatrix} A_x & A_y & A_z \\ B_x & B_y & B_z \\ C_x & C_y & C_z \end{vmatrix}$.

C. Guided Practice Problems

S 3.1 For the force vectors **P** and **Q**, where P = 860 N and Q = 500 N, determine
(a) the component of **P** in the direction of **Q**,
(b) the angle θ between **P** and **Q**,
(c) **R**, the cross product of **P** and **Q**.

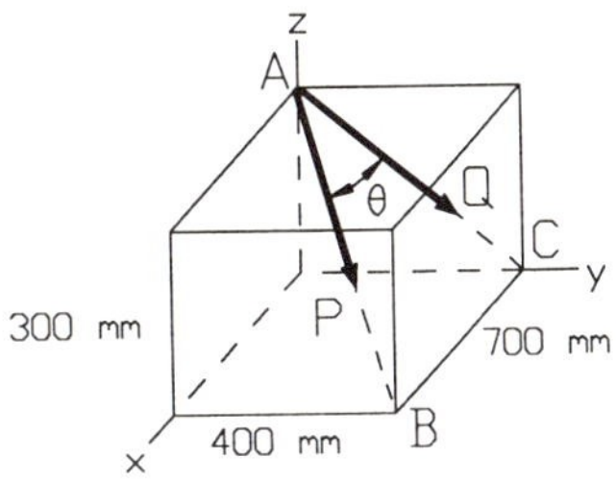

GUIDED SOLUTION

Part (a)

The component of **P** in the direction of **Q** is: $P\cos\theta = \mathbf{P} \bullet \boldsymbol{\lambda}_{AC}$, where $\boldsymbol{\lambda}_{AC}$ is the unit vector in the direction of **Q**.

1. Write **P** in vector form.

✍ (a) Write the vector from A to B: $\overrightarrow{AB} =$

✍ (b) Determine the unit vector from A toward B: $\boldsymbol{\lambda}_{AB} = \dfrac{\overrightarrow{AB}}{|\overrightarrow{AB}|} =$

✍ (c) Write **P** in vector form: $\mathbf{P} = P\boldsymbol{\lambda}_{AB} = 860\boldsymbol{\lambda}_{AB} =$

2. Write $\boldsymbol{\lambda}_{AC}$.

✍ (a) Write the vector from A to C: $\overrightarrow{AC} =$

✍ (b) Determine the unit vector from A toward C: $\boldsymbol{\lambda}_{AC} = \dfrac{\overrightarrow{AC}}{|\overrightarrow{AC}|} =$

3. Determine the component of **P** in the direction of **Q**.

✍ $P\cos\theta = \mathbf{P} \bullet \boldsymbol{\lambda}_{AC} =$

$P\cos\theta = 500\text{ N}$

Part (b)

1. Determine θ. Since the unit vectors in the direction of the forces have already been determined, $\cos\theta = \lambda_{AB} \bullet \lambda_{AC}$ will be used to determine θ.

✍ $\theta = \cos^{-1}(\lambda_{AB} \bullet \lambda_{AC}) =$

$\boxed{\theta = 54.5°}$

Part (c)

The cross product is: $\mathbf{R} = \mathbf{P} \times \mathbf{Q}$. Since we have already written **P** in vector form, we begin by writing **Q**.

1. Write **Q** in vector form.

✍ $\mathbf{Q} = Q\lambda_{AC} = 500\lambda_{AC} =$

2. Evaluate **R**.

✍ $$\mathbf{R} = \mathbf{P} \times \mathbf{Q} = \begin{vmatrix} \mathbf{i} & \mathbf{j} & \mathbf{k} \\ P_x & P_y & P_z \\ Q_x & Q_y & Q_z \end{vmatrix} = \begin{vmatrix} & & \\ & & \\ & & \end{vmatrix} =$$

$\boxed{\mathbf{R} = 210\mathbf{j} + 280\mathbf{k}\ \text{kN}^2}$

S 3.2 Determine the scalar triple product $\mathbf{A} \times \mathbf{B} \bullet \mathbf{C}$ for the vectors shown below. Use A = 65 lb, B = 25 lb, and C = 65 lb.

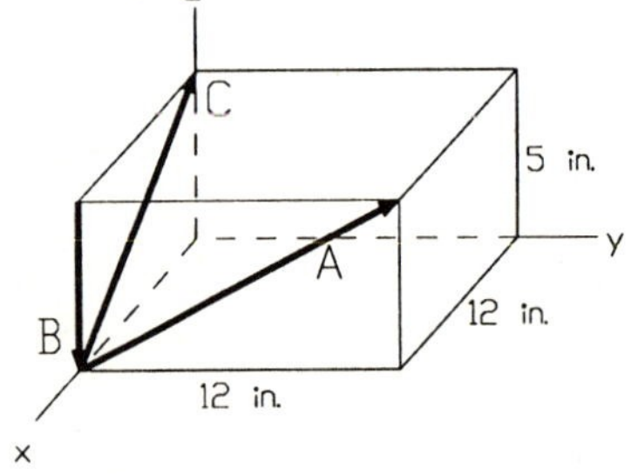

GUIDED SOLUTION

We begin by writing the vectors **A**, **B**, and **C** in rectangular form. Noting that the slopes of **A** and **B** are 5/12, their rectangular components can be conveniently found by trigonometry.

1. Write **A**, **B**, and **C** in vector form.

✍ $\mathbf{A} = A_y\mathbf{j} + A_z\mathbf{k} = (12/13)\,65\mathbf{j} + (5/13)\,65\mathbf{k} =$

$\mathbf{B} = -B\mathbf{k} =$

$\mathbf{C} = -C_x\mathbf{i} + C_z\mathbf{k} = -(12/13)\,65\mathbf{i} + (5/13)\,65\mathbf{k} =$

2. Evaluate the determinant.

✍ $$\mathbf{A}\times\mathbf{B}\bullet\mathbf{C} = \begin{vmatrix} A_x & A_y & A_z \\ B_x & B_y & B_z \\ C_x & C_y & C_z \end{vmatrix} = \begin{vmatrix} & & \\ & & \\ & & \end{vmatrix} =$$

$$\boxed{\mathbf{A}\times\mathbf{B}\bullet\mathbf{C} = 90 \times 10^3\ \text{lb}^3}$$

S 4. REDUCTION OF CONCURRENT FORCE SYSTEMS

Text Reference: Article 2.4; Sample Problems 2.1, 2.2

A. *You Should Understand:*

- The Principle of Transmissibility is used to move forces along their lines of action without changing their external effect on a rigid body.
- If forces $\mathbf{F}_1$, $\mathbf{F}_2$, ... are concurrent, their resultant force is $\mathbf{R} = \Sigma\,\mathbf{F} = \mathbf{F}_1 + \mathbf{F}_2 + \dots$.
- The resultant of concurrent forces passes through the point of concurrency.
- For a concurrent force system, three scalar equations are required to determine the components of **R**: $R_x = \Sigma\,F_x$, $R_y = \Sigma\,F_y$, $R_z = \Sigma\,F_z$.
- If the original forces lie in a plane, for example, the x-y plane, only two scalar equations are necessary to determine **R**: $R_x = \Sigma\,F_x$, $R_y = \Sigma\,F_y$.

B. *You Should be Able to:*

- Determine the resultant of a concurrent force system by summing the vectors: $\mathbf{R} = \mathbf{F}_1 + \mathbf{F}_2 + \dots$.
- Determine the resultant of a concurrent force system using scalar equations: $R_x = \Sigma\,F_x$, $R_y = \Sigma\,F_y$, $R_z = \Sigma\,F_z$, where R_x, R_y, and R_z are the rectangular components of **R**.

C. Guided Practice Problems

S 4.1 Determine the resultant **R** of the three forces shown. Use $F_1 = 250$ kN, $F_2 = 100$ kN, and $F_3 = 200$ kN.

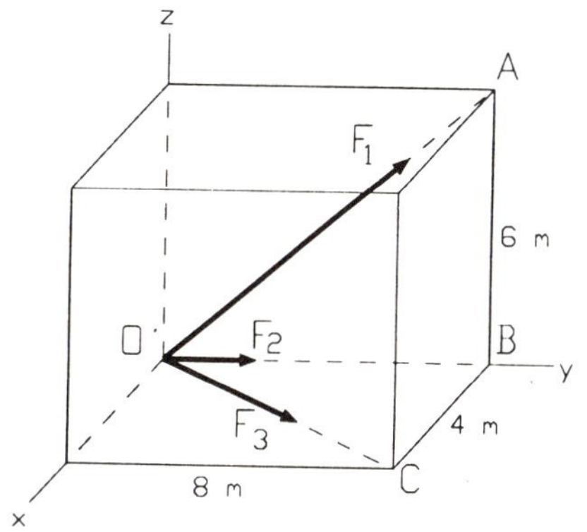

GUIDED SOLUTION

We begin by writing each force in vector form.

1. Write $\mathbf{F}_1$ in vector form.

✍ (a) Write the vector from O to A: $\overrightarrow{OA} =$

✍ (b) Determine the unit vector from O toward A: $\boldsymbol{\lambda}_{OA} = \dfrac{\overrightarrow{OA}}{|\overrightarrow{OA}|} =$

✍ (c) Write $\mathbf{F}_1$ in vector form: $\mathbf{F}_1 = F_1\boldsymbol{\lambda}_{OA} = 250\boldsymbol{\lambda}_{OA} =$

2. Write $\mathbf{F}_2$ in vector form.

✍ (a) Write the unit vector from O toward B: $\boldsymbol{\lambda}_{OB} =$

✍ (b) Write $\mathbf{F}_2$ in vector form: $\mathbf{F}_2 = F_2\boldsymbol{\lambda}_{OB} = 100\boldsymbol{\lambda}_{OB} =$

3. Write $\mathbf{F}_3$ in vector form.

✍ (a) Write the vector from O to C: $\overrightarrow{OC} =$

✍ (b) Determine the unit vector from O toward C: $\boldsymbol{\lambda}_{OC} = \dfrac{\overrightarrow{OC}}{|\overrightarrow{OC}|} =$

✍ (c) Write $\mathbf{F}_3$ in vector form: $\mathbf{F}_3 = F_3\boldsymbol{\lambda}_{OC} = 200\boldsymbol{\lambda}_{OC} =$

4. Determine the resultant by summing the vectors.

✍ $\mathbf{R} = \mathbf{F}_1 + \mathbf{F}_2 + \mathbf{F}_3 =$

R = 89.4**i** + 479**j** + 150**k** kN

S 4.2 Determine the resultant **R** of the force system shown.

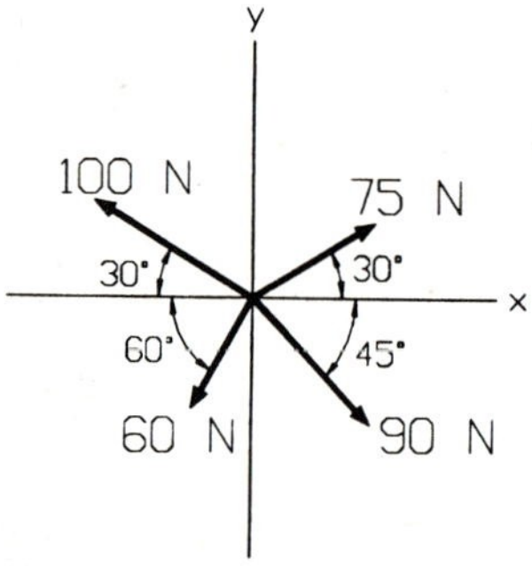

GUIDED SOLUTION

Since all of the forces lie in the x-y plane, it is convenient to determine the components of the vectors using trigonometry.

1. Determine R_x.

✍ $\overset{+}{\longrightarrow}$ $R_x = \Sigma F_x = 75 \cos 30° + 90 \cos 45° - 60 \cos 60° - 100 \cos 30° =$

2. Determine R_y.

✍ $+\uparrow$ $R_y = \Sigma F_y =$

3. Write **R**.

✍ $\mathbf{R} = R_x\mathbf{i} + R_y\mathbf{j} =$

R = 12.0**i** – 28.1**j** N

S 5. MOMENT OF A FORCE ABOUT A POINT

Text Reference: Article 2.5; Sample Problems 2.3, 2.4

A. *You Should Understand:*

- The moment of a force about a point is a measure of the tendency of a force to rotate a body about the point.

- One method of evaluating the moment of a force about a point O, $\mathbf{M}_O$, is the vector method: $\mathbf{M}_O = \mathbf{r} \times \mathbf{F}$, where **r** is the position vector from point O to any point on or along the line of action of **F**.

  Note: (i) The moment of a force about a point is a vector. The vector is perpendicular to both **r** and **F**, with its direction determined by the right-hand rule.

  (ii) The magnitude of $\mathbf{M}_O$ is $|\mathbf{M}_O| = |\mathbf{r} \times \mathbf{F}|$.

  (iii) Since $\mathbf{M}_O$ is independent of the point of application of **F** along its line of action, **F** may be considered a sliding vector for the purpose of determining its moment.

- Another method of evaluating the moment of a force about point O is the scalar method: $M_O = Fd$. The magnitude of **F** is F, and d, called the moment arm of the force, is the perpendicular distance from O to the line of action of **F**.

  Note: (i) The magnitude of $\mathbf{M}_O$ is $M_O = Fd$. However, the direction of $\mathbf{M}_O$ must be determined by inspection. Assume that a force is applied to a body that is fixed at a point O. The direction in which the body would tend to rotate is the direction of the moment about O. For example, in each of the two figures below, the forces act upward, have the same magnitude, and are the same perpendicular distance from point O. Therefore, in each case, $M_O = 500$ N•m. However, in (a) the direction of M_O is counterclockwise and in (b), the direction of M_O is clockwise.

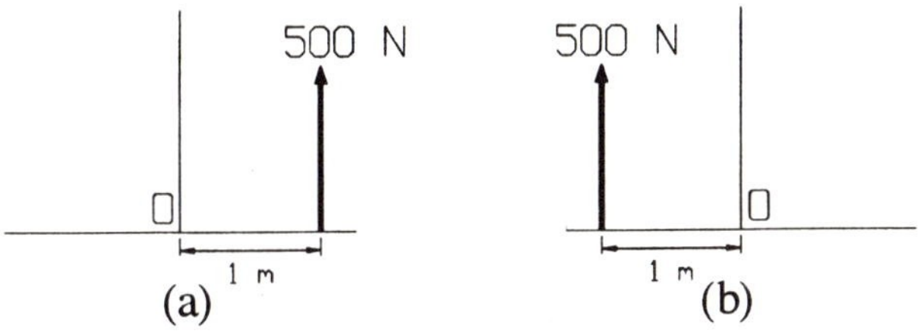

(a) (b)

  (ii) The scalar method is more convenient than the vector method when the moment arms are easy to determine.

- Principle of Moments: The sum of the moments of the components of a force about a point is equal to the moment of the resultant force about the point.

B. You Should be Able to:

- Write a position vector between any two points.
- Write a unit vector in any direction.
- Write a force in vector form.
- Calculate the moment of a force about a point O using the vector method:

$$\mathbf{M}_O = \mathbf{r} \times \mathbf{F} = \begin{vmatrix} \mathbf{i} & \mathbf{j} & \mathbf{k} \\ r_x & r_y & r_z \\ F_x & F_y & F_z \end{vmatrix}$$

Note: The steps involved in determining the moment of a force **F** about a point O are as follows:

1. Write **F** in vector form.
2. Choose a position vector **r** from point O to the line of action of **F** and write it in vector form.
3. Evaluate $\mathbf{M}_O = \mathbf{r} \times \mathbf{F}$.

- Calculate the moment of a force about a point using the scalar method: $M_O = Fd$, with the direction determined by inspection.
- Use the Principle of Moments to calculate the moment of a force about a point.

C. Guided Practice Problems

S 5.1 The 500-lb force acts at A directed toward B. Calculate the magnitude of the moment of this force about point C.

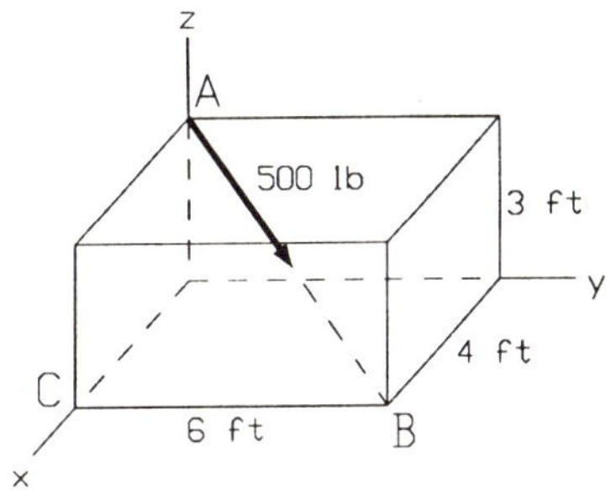

GUIDED SOLUTION

We must first decide whether to use the scalar method or the vector method. Since the perpendicular distance from C to the line of the force is not easy to find, we choose the vector method. We will evaluate $\mathbf{M}_C$ and then determine its magnitude.

1. Write **F** in vector form. We have $\mathbf{F} = 500\,\boldsymbol{\lambda}_{AB}$ since **F** acts along the line AB.

✍ (a) Write the vector from A to B: $\overrightarrow{AB} =$

✍ (b) Determine the unit vector from A toward B: $\boldsymbol{\lambda}_{AB} = \dfrac{\overrightarrow{AB}}{|\overrightarrow{AB}|} =$

✍ (c) Write **F** in vector form: $\mathbf{F} = F\boldsymbol{\lambda}_{AB} = 500\,\boldsymbol{\lambda}_{AB} =$

2. Choose **r** and write it in vector form. The position vector is a vector from point C to any point on the line of action of **F**. The obvious choices for the position vector are C to A and C to B. We choose the position vector from C to B.

✍ (a) Draw $\mathbf{r}_{CB}$ on the figure above.

✍ (b) Write $\mathbf{r}_{CB}$ in vector form: $\mathbf{r}_{CB} =$

3. Evaluate $\mathbf{M}_C$.

✍ $$\mathbf{M}_C = \mathbf{r}_{CB} \times \mathbf{F} = \begin{vmatrix} \mathbf{i} & \mathbf{j} & \mathbf{k} \\ & & \\ & & \end{vmatrix} =$$

4. Determine $|\mathbf{M}_C|$:

✍ $|\mathbf{M}_C| = \sqrt{M_{Cx}^2 + M_{Cy}^2 + M_{Cz}^2} =$

$|\mathbf{M}_C| = 1920$ lb•ft

S 5.2 Using the scalar method, calculate the moment of the 390-N force **F** about point C by the following methods:
(a) Locating the rectangular components of **F** at point A.
(b) Locating the rectangular components of **F** at point B.
(c) Using Fd.

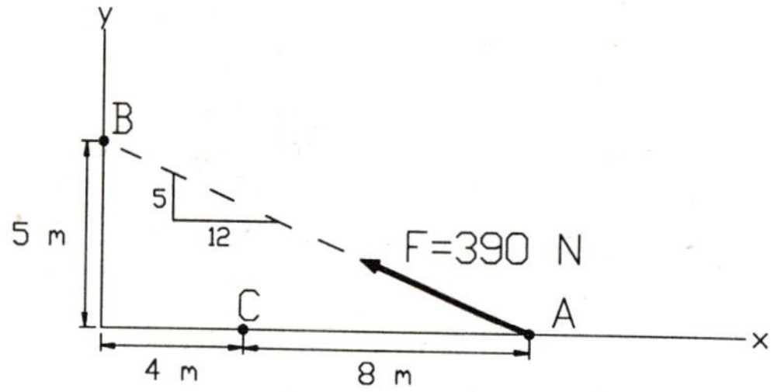

Guided Solution

Part (a)

Perpendicular distances (moment arms) are needed from point C to each of the rectangular components of **F** located at A. Note the 5/12 slope of AB.

1. Determine the magnitude of the vertical and horizontal components of **F**.

✍ $F_y = (5/13)F =$

$F_x = (12/13)F =$

2. Show F_x and F_y acting at A on the sketch.

✍

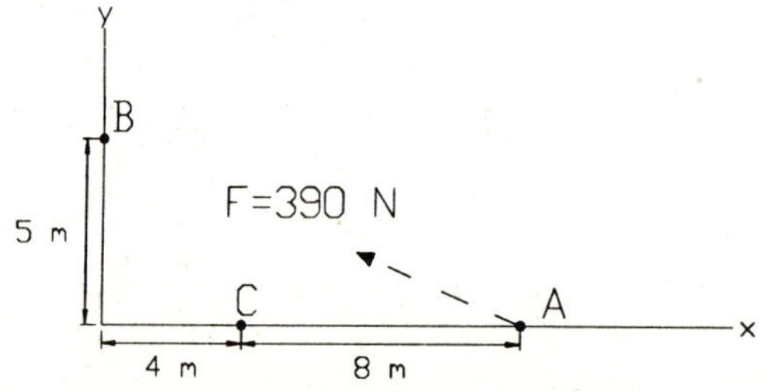

3. <u>Determine the perpendicular distance from C to each of the two components of</u> **F**:

✍ Perpendicular distance from C to line of F_y =

Perpendicular distance from C to line of F_x =

4. <u>Calculate M_C</u> (we choose CCW to be positive):

✍ ↺+ $M_C = \Sigma(Fd) =$

Note: The answer must <u>always</u> specify the direction of the moment (CW or CCW).

M_C = 1200 N•m CCW

Part (b)

This part of the problem is to be done in a manner identical to part (a) above, with the exception that the components of **F** are to be drawn at point B. (Remember that a force is a sliding vector and can be drawn in its component form anywhere along its line of action <u>without</u> changing its moment.)

1. <u>On the sketch below, show F_x and F_y acting at B</u>.

✍

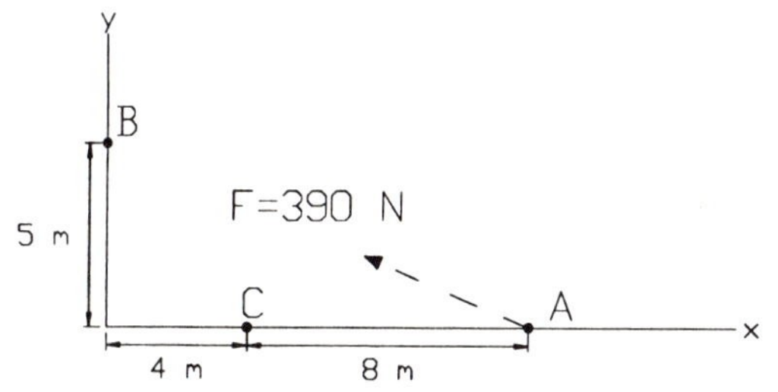

2. <u>Determine the perpendicular distance from C to each of the two components</u>:

✍ Perpendicular distance from C to line of F_y =

Perpendicular distance from C to line of F_x =

3. <u>Calculate M_C</u> (we choose CCW to be positive):

✍ ↺+ $M_C = \Sigma(Fd) =$

M_C = 1200 N•m CCW

Part (c)

For this part of the problem, we must find the perpendicular distance, d, from C to the line of action of **F**.

1. Determine d. From C we draw a line that is perpendicular to the line of action of **F** (i.e., the line AB). The right triangle CDA is formed, as shown below. Note that $\sin\theta = d/8$.

✍

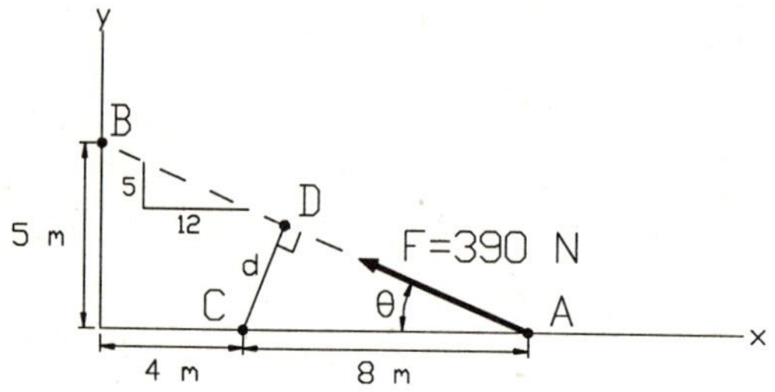

$d = 8\ \sin\theta = 8(5/13) =$

2. Calculate Fd and determine the direction of the moment.

✍ $M_C = Fd =$

M_C = 1200 N•m CCW

S 6. Moment of a Force about an Axis

Text Reference: Article 2.6; Sample Problems 2.5, 2.6

A. You Should Understand:

- The moment of a force about an axis is a vector that is a measure of the tendency of a force to rotate a body about that axis.

- A force that is parallel to an axis has no moment about that axis.

- A force that intersects an axis has no moment about that axis.

- One method of evaluating the moment of a force about an arbitrary a-a axis is the vector method: $\mathbf{M}_{a\text{-}a} = (\mathbf{M}_O \cdot \boldsymbol{\lambda}_{a\text{-}a})\,\boldsymbol{\lambda}_{a\text{-}a} = [(\mathbf{r} \times \mathbf{F}) \cdot \boldsymbol{\lambda}_{a\text{-}a}]\,\boldsymbol{\lambda}_{a\text{-}a}$, where:

  $\mathbf{M}_O$ is the moment about an arbitrary point O on the a-a axis,

  $\mathbf{r}$ is the position vector from point O to any point on the line of action of $\mathbf{F}$,

  $\boldsymbol{\lambda}_{a\text{-}a}$ is the unit vector in the direction of the a-a axis.

  Note: (i) The magnitude of $\mathbf{M}_{a\text{-}a}$ is $|\mathbf{M}_{a\text{-}a}| = \mathbf{M}_O \cdot \boldsymbol{\lambda}_{a\text{-}a} = (\mathbf{r} \times \mathbf{F}) \cdot \boldsymbol{\lambda}_{a\text{-}a}$

  (ii) The moment of a force about the a-a axis is the component of $\mathbf{M}_O$ in the direction of the a-a axis, where O is any point on the a-a axis.

  (iii) The direction in which the unit vector is taken determines the positive direction for $\mathbf{M}_{a\text{-}a}$. If $|\mathbf{M}_{a\text{-}a}|$ is positive, the direction of $\mathbf{M}_{a\text{-}a}$ is the same as $\boldsymbol{\lambda}_{a\text{-}a}$. If $|\mathbf{M}_{a\text{-}a}|$ is negative, the direction of $\mathbf{M}_{a\text{-}a}$ is opposite of $\boldsymbol{\lambda}_{a\text{-}a}$.

  (iv) If point O is the origin of a rectangular coordinate system, then $\mathbf{M}_O = M_x\mathbf{i} + M_y\mathbf{j} + M_z\mathbf{k}$, where M_x, M_y, and M_z are the moments about the x, y, and z axes, respectively.

- A second method of evaluating the moment of a force about the a-a axis is the scalar method. If a force $\mathbf{F}$ lies in a plane that is perpendicular to the a-a axis, its moment is $M_{a\text{-}a} = Fd$. The magnitude of $\mathbf{F}$ is F, and d is the length of the moment arm that lies in the plane of $\mathbf{F}$, perpendicular to the line of action of $\mathbf{F}$.

  Note: (i) The direction in which a force would tend to rotate a body about the a-a axis can be determined in the following way. Assume that a force is applied to a body in which the a-a axis is fixed. The direction in which the body would tend to rotate is the direction of the moment about the a-a axis.

  (ii) The magnitude of the moment of the force $\mathbf{F}$ about the a-a axis is $M_{a\text{-}a} = Fd$. The direction of the vector $\mathbf{M}_{a\text{-}a}$ must be determined by applying the right hand rule to the direction of rotational tendency of the force.

(iii) The scalar method is convenient when the moment arm of the force is easy to determine.

- Principle of Moments: The sum of the moments of the components of a force about an axis is equal to the moment of the resultant force about the axis.

B. *You Should be Able to:*

- Determine the moments about the x, y, and z axes from $\mathbf{M}_O$, using $\mathbf{M}_O = M_x\mathbf{i} + M_y\mathbf{j} + M_z\mathbf{k}$, where O is the origin of a rectangular coordinate system.

- Calculate the magnitude of the moment of a force about the a-a axis using the vector method:

$$|\mathbf{M}_{a\text{-}a}| = \mathbf{M}_O \cdot \lambda_{a\text{-}a}, \text{ where } \mathbf{M}_O = \mathbf{r} \times \mathbf{F} = \begin{vmatrix} \mathbf{i} & \mathbf{j} & \mathbf{k} \\ r_x & r_y & r_z \\ F_x & F_y & F_z \end{vmatrix}.$$

These two calculations can be combined into one step using the scalar triple product:

$$|\mathbf{M}_{a\text{-}a}| = \mathbf{M}_O \cdot \lambda_{a\text{-}a} = (\mathbf{r} \times \mathbf{F}) \cdot \lambda_{a\text{-}a} = \begin{vmatrix} r_x & r_y & r_z \\ F_x & F_y & F_z \\ \lambda_x & \lambda_y & \lambda_z \end{vmatrix}.$$

Note: The steps involved in determining the moment of a force **F** about the a-a axis are as follows:

1. Write **F** in vector form.

2. Choose a position vector **r**, from the a-a axis to the line of action of **F**, and write it in vector form.

3. Write the unit vector, $\lambda_{a\text{-}a}$.

4. Evaluate the magnitude of $\mathbf{M}_{a\text{-}a}$: $|\mathbf{M}_{a\text{-}a}| = \mathbf{M}_O \cdot \lambda_{a\text{-}a} = (\mathbf{r} \times \mathbf{F}) \cdot \lambda_{a\text{-}a}$.

5. Write $\mathbf{M}_{a\text{-}a}$ in vector form: $\mathbf{M}_{a\text{-}a} = |\mathbf{M}_{a\text{-}a}|\, \lambda_{a\text{-}a}$

- Determine the magnitude and direction of the moment of a force about an axis using the scalar method: $M_{a\text{-}a} = Fd$. The direction of the moment must be determined by inspection.

- Use the Principle of Moments to calculate the moment of a force about an axis.

C. Guided Practice Problems

S 6.1 The 170-kN force **P** acts at A directed toward B. Determine the moment of **P** about an axis directed from C to D.

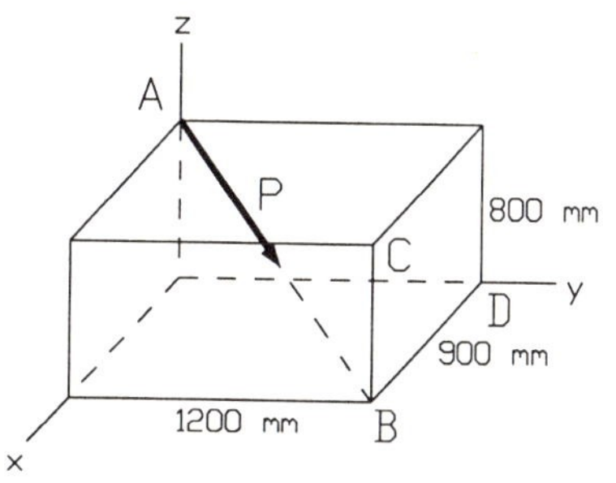

GUIDED SOLUTION

First, we must decide whether to use the scalar or vector method. Since the moment arms are not easy to find, we choose the vector method.

1. Write **P** in vector form.

✍ (a) Write the vector from A to B: $\overrightarrow{AB}$ =

✍ (b) Evaluate the unit vector from A to B: $\lambda_{AB} = \dfrac{\overrightarrow{AB}}{|\overrightarrow{AB}|} =$

✍ (c) Write **P** in vector form: $\mathbf{P} = P\lambda_{AB} = 170\,\lambda_{AB} =$

2. Choose **r** and write it in vector form. The position vector **r** is a vector from any point on CD to any point along the line of **P**. The obvious choices for **r** are as follows: from C to A, from C to B, from D to B, or from D to A. We choose the position vector from C to B.

✍ (a) Draw $\mathbf{r}_{CB}$ on the figure above.

✍ (b) Write $\mathbf{r}_{CB}$ in vector form: $\mathbf{r}_{CB}$ =

3. Write λ_{CD} in vector form. The problem states that the axis is to be directed from C to D. Therefore, the unit vector is to be written in the direction from C toward D.

✍ $\lambda_{CD} = \dfrac{\overrightarrow{CD}}{|\overrightarrow{CD}|} =$

4. Evaluate $|\mathbf{M}_{CD}|$. Use the scalar triple product:

✍ $|\mathbf{M}_{CD}| = (\mathbf{r} \times \mathbf{P}) \cdot \boldsymbol{\lambda}_{CD} = \begin{vmatrix} & & \\ & & \\ & & \end{vmatrix} =$

5. Write $\mathbf{M}_{CD}$.

✍ $\mathbf{M}_{CD} = |\mathbf{M}_{CD}| \boldsymbol{\lambda}_{CD} =$

$\mathbf{M}_{CD} = 53.6\,\mathbf{i} + 47.7\,\mathbf{k}$ kN•m

S 6.2 Using the scalar method, calculate the moment of the 390-lb force **F** about the coordinate (x,y,z) axes and about the a-a axis.

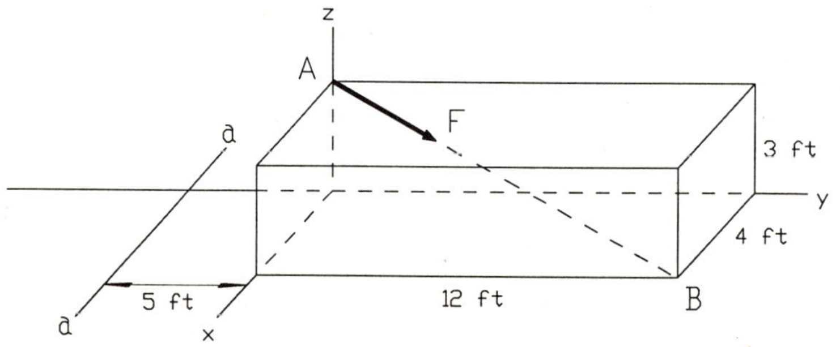

GUIDED SOLUTION

The rectangular components of **F** (F_x, F_y, F_z) and their moment arms from the x, y, z, and a-a axes must be determined. One way to determine the rectangular components is to write **F** in vector form. Another way is to use trigonometry by using the angles that **F** makes with each of the coordinate axes. Since the angles are not easy to determine, we write **F** in vector form.

1. Write **F** in vector form.

✍ (a) Write the vector from A to B: $\overrightarrow{AB} =$

✍ (b) Evaluate the unit vector from A to B: $\boldsymbol{\lambda}_{AB} = \dfrac{\overrightarrow{AB}}{|\overrightarrow{AB}|} =$

✍ (c) Write **F** in vector form: $\mathbf{F} = F\boldsymbol{\lambda}_{AB} = 390\,\boldsymbol{\lambda}_{AB} =$

2. Determine the rectangular components of **F** (from the results of step 1 above).

$F_x =$ $F_y =$ $F_z =$

Show F_x, F_y, and F_z acting at A on the figure above:

3. Calculate M_x and determine its direction.

(a) F_x has no moment about the x-axis because ______________________

F_y has a moment about the x-axis. Its moment arm is d =

F_z has no moment about the x-axis because ______________________

(b) Calculate M_x: $M_x = \Sigma(Fd) =$

4. Calculate M_y and determine its direction.

(a) F_x has a moment about the y-axis. Its moment arm is d =

F_y has no moment about the y-axis because ______________________

F_z has no moment about the y-axis because ______________________

(b) Calculate M_y: $M_y = \Sigma(Fd) =$

5. Calculate M_z and determine its direction:

(a) F_x has no moment about the z-axis because ______________________

F_y has no moment about the z-axis because ______________________

F_z has no moment about the z-axis because ______________________

(b) Calculate M_z: $M_z = \Sigma(Fd) =$

6. Calculate $M_{a\text{-}a}$ and determine its direction.

(a) F_x has no moment about the a-a axis because ______________________

F_y has a moment about the a-a axis. Its moment arm is $d_1 =$

F_z has a moment about the a-a axis. Its moment arm is $d_2 =$

(b) Calculate $M_{a\text{-}a}$: $M_{a\text{-}a} = \Sigma(Fd) =$

$M_x = -1080$ lb•ft
$M_y = 360$ lb•ft
$M_z = 0$
$M_{a\text{-}a} = -1530$ lb•ft

S 7. COUPLES

Text Reference: Article 2.7; Sample Problems 2.7 - 2.9

. *You Should Understand:*

- The term *couple* refers to two forces that are equal in magnitude, but are oppositely directed along parallel, non-collinear lines of action.

- The moment of a couple can be calculated by summing the moments of the two forces that form the couple.

- A couple can rotate a body since it has a moment, but it cannot translate a body since the resultant of the two forces is zero.

- Couples are important because of the following property: The moment of a couple is the same about every point.

- The moment of a couple is a free vector, referred to as the couple-vector.

- The moment of a couple can be calculated using **C** = **r** x **F** or C = Fd. When using C = Fd, the direction of the moment of the couple must be determined by inspection. The direction in which a body would tend to rotate if the couple were applied is the direction of the moment of the couple.

- The terms *couple* and *moment of a couple* are often used synonymously.

- A couple may be represented in one of three ways. For example, the following are equivalent representations of a 500-N•m counterclockwise couple lying in the y-z plane.

    1. Two forces that are equal in magnitude but are oppositely directed along parallel, non-collinear lines of action:

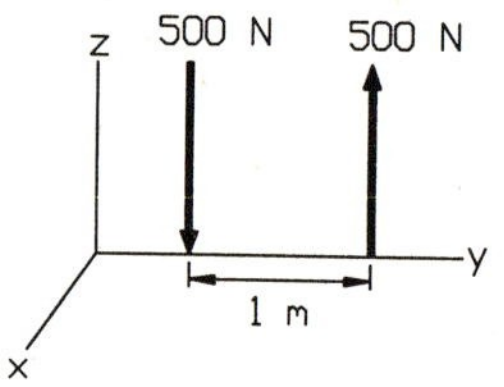

or

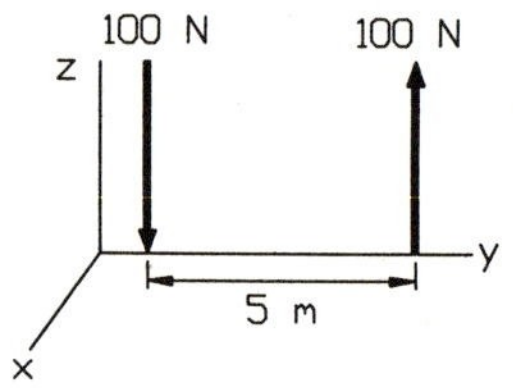

2. The magnitude of the couple stated, the direction of its moment indicated by a counterclockwise arrow in the y-z plane:

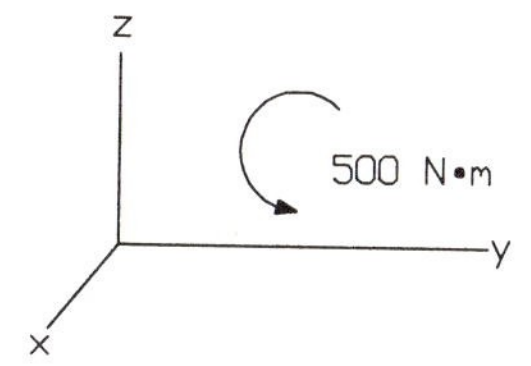

3. A couple-vector, $\mathbf{C} = 500\mathbf{i}$ N•m (using the right-hand rule):

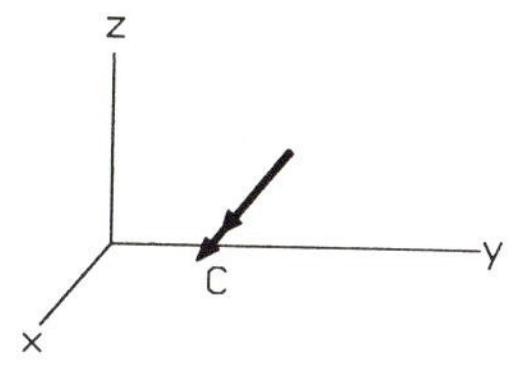

- For a given couple **C**, there is no point for which the moment of the couple is zero.

- The moment of a couple **C** about an a-a axis is: $\mathbf{M}_{a\text{-}a} = \mathbf{C} \cdot \lambda_{a\text{-}a}$, where $\lambda_{a\text{-}a}$ is a unit vector in the direction of the a-a axis.

  Note: $\mathbf{M}_{a\text{-}a}$ is the component of **C** in the direction of the a-a axis.

- The moment of a couple **C** is zero about an axis that is perpendicular to the vector **C**. (The axis lies in the plane of the couple.)

B. You Should be Able to:

- Determine a couple by summing the moments of the two forces that comprise it about any point.

- Determine the moment of a couple about an axis.

C. Guided Practice Problem

S 7.1 The two 100-N forces shown form a couple.
(a) Write the couple-vector.
(b) Determine the moment of the couple about point O.
(c) Determine the moment of the couple about an axis from O toward E and express your answer in vector form.

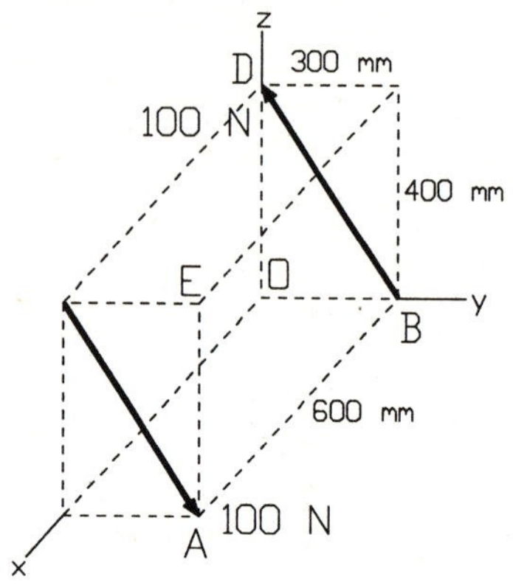

GUIDED SOLUTION

Part (a)

The couple-vector can be found by summing the moments of the two forces about a convenient point. We choose point A because one of the forces passes through point A and therefore, has no moment about it. (Points B and D could have also been chosen for the same reason.)

1. Choose $\mathbf{r}$ and write it in vector form. Since we are summing moments about A, we choose $\mathbf{r}$ to be $\mathbf{r}_{AB}$.

✍ (a) Draw $\mathbf{r}_{AB}$ on the figure above.

✍ (b) Write $\mathbf{r}_{AB}$: $\mathbf{r}_{AB} =$

2. Write $\mathbf{F}$ in vector form. Since we are taking the moment about A, the force passing through B must be written in vector form: $\mathbf{F} = 100\boldsymbol{\lambda}_{BD}$.

✍ (a) Write $\boldsymbol{\lambda}_{BD}$: $\boldsymbol{\lambda}_{BD} =$

✍ (b) Write $\mathbf{F}$: $\mathbf{F} = 100\boldsymbol{\lambda}_{BD} =$

3. Determine **C**.

✍ $\mathbf{C} = \mathbf{M}_A = \mathbf{r}_{AB} \times \mathbf{F} = \begin{vmatrix} \mathbf{i} & \mathbf{j} & \mathbf{k} \\ & & \\ & & \end{vmatrix} =$

$\mathbf{C} = 48.0\mathbf{j} + 36.0\mathbf{k}$ N•m

Part (b)

1. Determine the moment of the couple about point O. Since the moment of a couple is the same about any point, we have:

✍ $\mathbf{M}_O = \mathbf{M}_A = \mathbf{C} =$

$\mathbf{M}_O = 48.0\mathbf{j} + 36.0\mathbf{k}$ N•m

Part (c)

The moment of the couple about the axis from O to E is the component of the couple in the direction of the axis.

1. Write the unit vector in the direction from O toward E.

✍ $\lambda_{OE} =$

2. Determine $\mathbf{M}_{OE}$.

✍ (a) Calculate M_{OE}: $M_{OE} = \mathbf{C} \cdot \lambda_{OE} =$

✍ (b) Write $\mathbf{M}_{OE}$: $\mathbf{M}_{OE} = M_{OE}\lambda_{OE} =$

$\mathbf{M}_{OE} = 28.3\mathbf{i} + 14.2\mathbf{j} + 18.9\mathbf{k}$ N•m

S 8. REPLACING A FORCE WITH A FORCE AND A COUPLE

Text Reference: Article 2.8; Sample Problems 2.10, 2.11

A. *You Should Understand:*

- A force **F** acting at point A can be replaced by the force **F** acting at point B and a couple of transfer $\mathbf{C}^T$.

- When a force **F** is moved from point A to point B, the couple of transfer, $\mathbf{C}^T$, equals the moment of **F** about point B. Therefore, the plane of the couple of transfer contains the original force **F** and the new point B, and the couple-vector $\mathbf{C}^T$ is perpendicular to the vector **F**. As is the case for all couples, $\mathbf{C}^T$ is a free vector.

B. *You Should be Able to:*

- Replace a force acting at a point by a force acting at another point and a couple of transfer.

C. Guided Practice Problems

S 8.1 A force **F** and a couple **C** act at point A as shown. Find the equivalent force-couple system acting at point D. Use F = 60 lb and C = 180 lb•in.

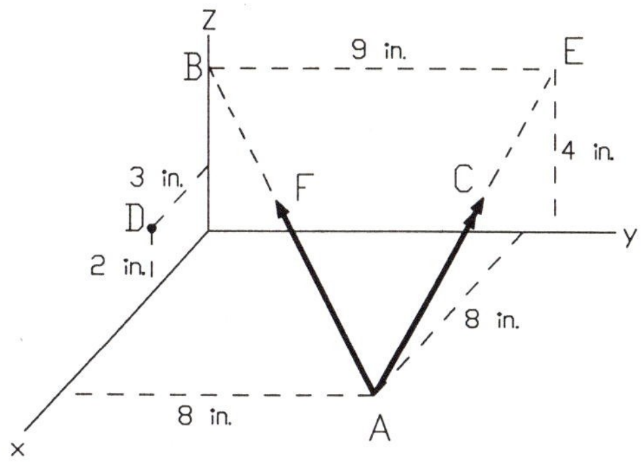

GUIDED SOLUTION

The couple **C** is a free vector and can be moved anywhere without changing its effect. A couple of transfer is needed when **F** is moved to point D. The original couple **C** and the couple of transfer can then be summed to determine the resultant couple.

1. <u>Write</u> **F** <u>in vector form.</u>

✍ (a) Write the unit vector from A toward B: $\lambda_{AB} = \dfrac{\overrightarrow{AB}}{|\overrightarrow{AB}|} =$

✍ (b) Write **F**: $\mathbf{F} = 60\lambda_{AB} =$

2. <u>Determine the couple of transfer</u>, $\mathbf{C}^T$. The couple of transfer equals the moment of **F** about D. A vector from point D to a point on the line AB needs to be chosen. We choose the vector from D to B.

✍ (a) Write $\mathbf{r}_{DB}$: $\mathbf{r}_{DB} =$

✍ (b) Determine $\mathbf{C}^T$:

$$\mathbf{C}^T = \mathbf{M}_D = \mathbf{r}_{DB} \times \mathbf{F} = \begin{vmatrix} \mathbf{i} & \mathbf{j} & \mathbf{k} \\ & & \\ & & \end{vmatrix} =$$

3. Write **C** in vector form.

✍ (a) Write the unit vector from A toward E: $\boldsymbol{\lambda}_{AE} = \dfrac{\overrightarrow{AE}}{|\overrightarrow{AE}|} =$

✍ (b) Write **C**: $\mathbf{C} = 180\boldsymbol{\lambda}_{AE} =$

4. Add the two couples.

✍ $\mathbf{C}^R = \mathbf{C} + \mathbf{C}^T =$

5. Write the equivalent force-couple system acting at D.

✍ $\mathbf{F} =$

$\mathbf{C}^R =$

$\mathbf{F} = -40.0\mathbf{i} - 40.0\mathbf{j} + 20.0\mathbf{k}$ lb
$\mathbf{C}^R = -80.0\mathbf{i} + 200\mathbf{k}$ lb•in

S 9. REDUCING A FORCE SYSTEM TO A FORCE AND A COUPLE

Text Reference: Article 3.2; Sample Problems 3.1, 3.2

A. *You Should Understand:*

- A force system can be reduced to a resultant force **R** at any point (e.g. point O) and a resultant couple $\mathbf{C}^R$. The resultant force-couple system has the same external effect on a rigid body as the original system.

- The resultant force, **R**, acting at point O is given by: $\mathbf{R} = \Sigma\mathbf{F} = \mathbf{F}_1 + \mathbf{F}_2 + \dots$.

- The resultant couple, $\mathbf{C}^R$, is given by: $\mathbf{C}^R = \Sigma\mathbf{M}_O$. (If the original force system includes couples, they are added to the couples of transfer to form the resultant couple.)

- The rectangular components of **R** are: $R_x = \Sigma F_x$, $R_y = \Sigma F_y$, $R_z = \Sigma F_z$. The rectangular components of $\mathbf{C}^R$ are: $C^R{}_x = \Sigma M_x$, $C^R{}_y = \Sigma M_y$, $C^R{}_z = \Sigma M_z$, where M_x, M_y, and M_z are moments about the x-, y-, and z-axes, respectively.

  Note: If the forces act in a plane, for example, the x-y plane, the scalar equations necessary to determine the resultant force-couple system are: $R_x = \Sigma F_x$, $R_y = \Sigma F_y$, and $C^R = \Sigma M_O$

- $\mathbf{C}^R$ is generally not perpendicular to **R**.

B. *You Should be Able to:*

- Replace an arbitrary force system by an equivalent force-couple system acting at a given point.

  Note: The steps involved in reducing a force system consisting of forces $\mathbf{F}_1$, $\mathbf{F}_2$, and $\mathbf{F}_3$ to an equivalent force-couple system acting at point O are as follows:

  1. Add the forces to determine the resultant force **R**: $\mathbf{R} = \mathbf{F}_1 + \mathbf{F}_2 + \mathbf{F}_3$.

  2. Show **R** acting at point O.

  3. Determine the resultant couple $\mathbf{C}^R$ by summing the moments of the original force system about point O: $\mathbf{C}^R = \Sigma\mathbf{M}_O$.

C. Guided Practice Problems

S 9.1 The system shown consists of three forces and a couple. Reduce this system to a force-couple system acting at point O. Show the results on a sketch of the coordinate system.

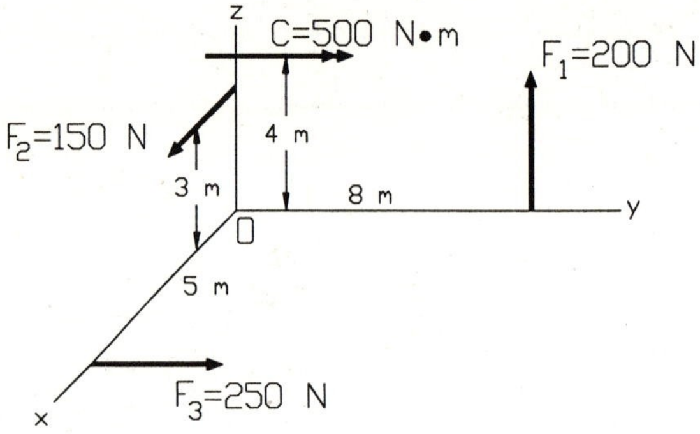

GUIDED SOLUTION

Note that the given couple will not affect the magnitude of the resultant force ($\mathbf{R} = \Sigma\mathbf{F}$), but it does affect the resultant couple ($\mathbf{C}^R = \Sigma\mathbf{M}_O$).

1. Determine the resultant force **R**.

✍ (a) Write each force in vector form:

$\mathbf{F}_1 =$

$\mathbf{F}_2 =$

$\mathbf{F}_3 =$

✍ (b) Add the forces: $\mathbf{R} = \Sigma\mathbf{F} =$

2. Sketch **R** at O.

✍

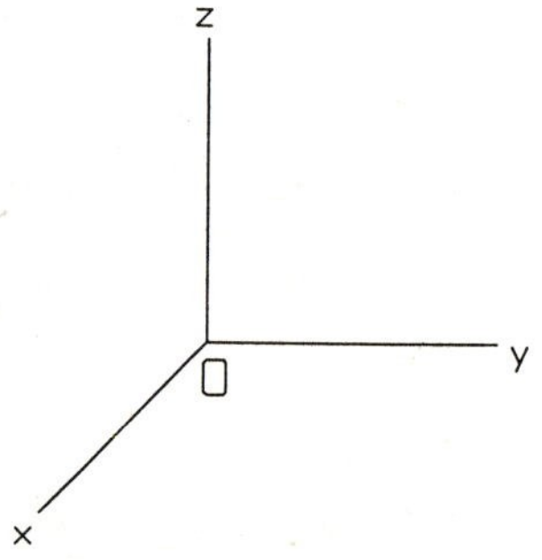

3. Find the resultant couple. The resultant couple is found by summing moments about O ($\mathbf{C}^R = \Sigma \mathbf{M}_O$). Use the right hand rule to determine whether the direction of the moments is in the positive or negative coordinate (x, y, or z) direction.

✍ For $\mathbf{F}_1$: $\mathbf{M}_O =$

For $\mathbf{F}_2$: $\mathbf{M}_O =$

For $\mathbf{F}_3$: $\mathbf{M}_O =$

For $\mathbf{C}$: $\mathbf{M}_O = \mathbf{C} =$

$\mathbf{C}^R = \Sigma \mathbf{M}_O =$

4. Reduce the resultant to the force-couple system at O. Write the resultant couple-vector at O and show the vectors on a sketch of coordinate system.

✍ $\mathbf{R} =$

$\mathbf{C}^R =$

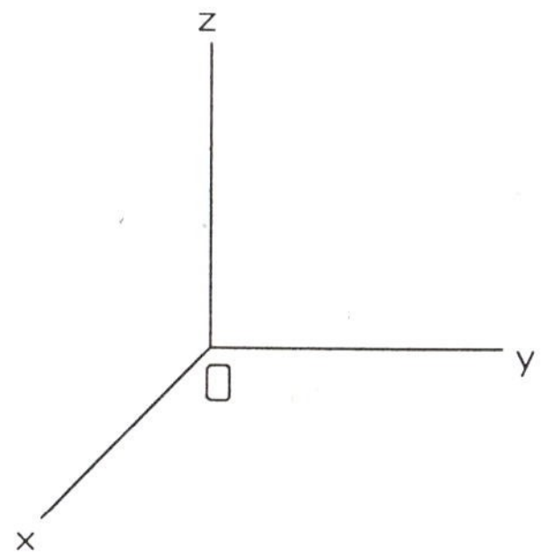

$\mathbf{R} = 150\mathbf{i} + 250\mathbf{j} + 200\mathbf{k}$ N
$\mathbf{C}^R = 1600\mathbf{i} + 950\mathbf{j} + 1250\mathbf{k}$ N•m

S 10. RESULTANTS OF COPLANAR FORCE SYSTEMS

Text Reference: Articles 3.3, 3.4; Sample Problems 3.3 -3.6

A. You Should Understand:

- The resultant of a force system is the "simplest system that can replace the original system without changing the external effect on a body."
- The resultant of a force system may be a force, a couple, or the combination of a force and a couple.
- The terms *resultant of a force system*, *resultant force*, and *resultant couple* have different meanings, as described in Article 3.3 of the text.
- The resultant of a general coplanar force system is either a force or a couple. The resultant cannot be a force and a couple because a force and a couple can always be reduced to a single force in the coplanar case.
- The resultant of a concurrent, coplanar force system is a force that passes through point of concurrency.
- The resultant of a parallel, coplanar force system, is a force or a couple.
- The equations that determine the resultant of a force system depend on the type of force system. The following table summarizes the equations used to determine the resultant of coplanar force systems.

Resultants of Coplanar Force Systems

Force System	Resultant of System	Equations that Determine the Resultant
Concurrent (at point O)	A force, **R**, through point O.	$\mathbf{R} = \Sigma \mathbf{F}$ $(R_x = \Sigma F_x, R_y = \Sigma F_y)$
Parallel (to y-axis)	A force, **R** (parallel to y-axis), or a couple, $\mathbf{C}^R$ ($\mathbf{C}^R$ has no y-component).	$R = \Sigma F_y$ The perpendicular distance from point O (O is any point) to the line of action of **R** is $d = \Sigma M_O / R$.
		If $\Sigma F_y = 0$, and $\Sigma M_O \neq 0$, the resultant is a couple: $C^R = \Sigma M_O$.
General	A force, **R**, or a couple, $\mathbf{C}^R$.	$\mathbf{R} = \Sigma \mathbf{F}$ $(R_x = \Sigma F_x, R_y = \Sigma F_y)$ The perpendicular distance from any point O to the line of action of **R** is $d = \Sigma M_O / R$.
		If $\mathbf{R} = 0$ $(\Sigma F_x = 0$ and $\Sigma F_y = 0)$, and $\Sigma M_O \neq 0$, the resultant is a couple: $C^R = \Sigma M_O$.

B. You Should be Able to:

- Determine the resultant of coplanar force systems.

C. Guided Practice Problems

S 10.1 Determine the resultant of the force system shown that consists of three forces and one couple. If the resultant is a force, locate it with respect to point O. Show your answer on a sketch of the coordinate system.

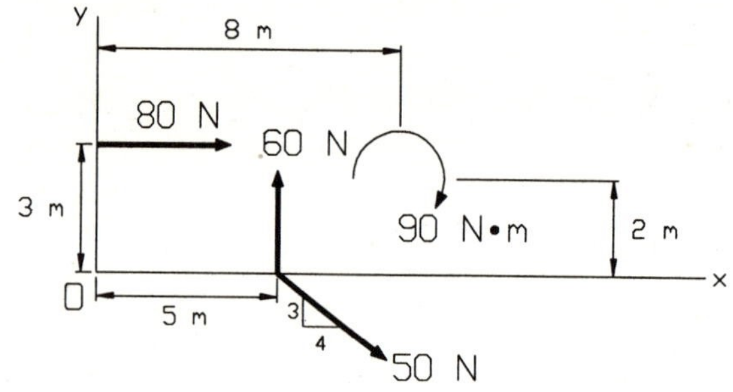

GUIDED SOLUTION

1. Determine the resultant. This is a general coplanar force system.

(a) Sum the forces in the x and y directions.

$R_x = \Sigma F_x$: $\overset{+}{\rightarrow}$ $R_x =$

$R_y = \Sigma F_y$: $+\uparrow$ $R_y =$

Since $\Sigma \mathbf{F} \neq 0$, the resultant is a force.

(b) Calculate the magnitude of **R**: $R = \sqrt{R_x^{\,2} + R_y^{\,2}} =$

Label the slope of the vector **R**:

2. Locate the line of action of **R** with respect to point O. We choose point O as the moment center and determine the perpendicular distance d from O to the line of **R**.

(a) Sum of the moments about O. (Recall that the moment of a couple is the same about every point.)

$\circlearrowleft +$ $\Sigma M_O =$

(b) Calculate the perpendicular distance d: $d = \Sigma M_O / R =$

3. Show the resultant of the force system on a sketch of the coordinate system.

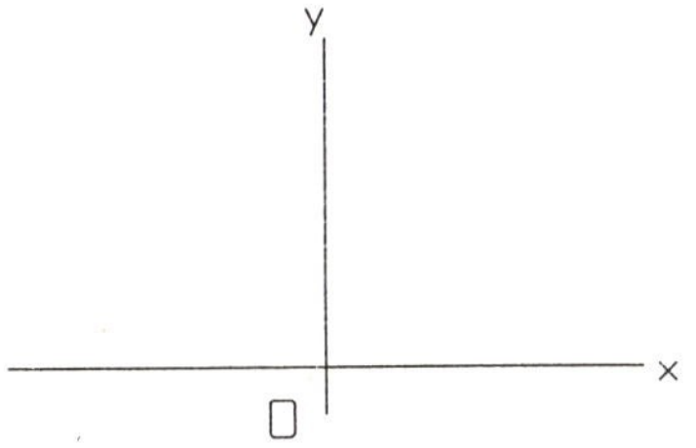

On the sketch:
R = 124 N at 1/4 slope
d = 1.45 m (refer to the solution for the slope)

S 10.2 Determine the resultant of the force system shown that consists of three forces. If the resultant is a force, locate it with respect to point O. Show your answer on a sketch of the coordinate system.

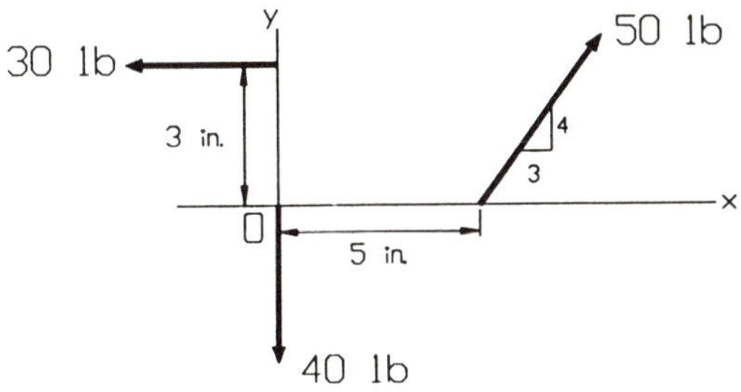

GUIDED SOLUTION

1. Determine the resultant.

(a) Sum the forces in the x and y directions.

$R_x = \Sigma F_x$: $\quad \overset{+}{\rightarrow} \quad R_x =$

$R_y = \Sigma F_y$: $\quad +\uparrow \quad R_y =$

Since $\Sigma \mathbf{F} = 0$, the resultant is not a force. It may be a couple.

(b) Sum the moments about point O. (Point O is selected for convenience.)

$\circlearrowleft + \quad \Sigma M_O =$

2. Show the resultant of the force system on a sketch of the coordinate system.

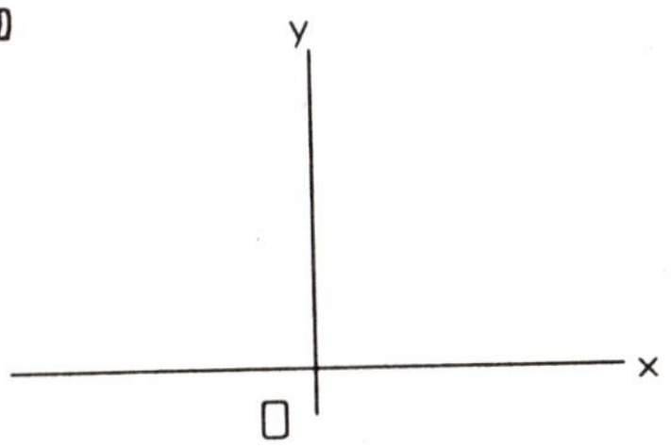

On the sketch: C = 290 lb•in CCW, anywhere

S 11. RESULTANTS OF NONCOPLANAR FORCE SYSTEMS

Text Reference: Article 3.5; Sample Problems 3.7 - 3.10

A. *You Should Understand:*

- The resultant of a concurrent, noncoplanar force system is a force **R** acting through the point of concurrency.

- The resultant of a parallel, noncoplanar force system is a force **R** or a couple $\mathbf{C}^R$.

- If **R** and $\mathbf{C}^R$ are not perpendicular, the resultant may be reduced to a wrench, a collinear force-couple system. The line formed by the collinear force-couple system is called the axis of the wrench.

- The equations that determine the resultant of a force system depend on the type of force system. The following table summarizes the equations used to determine the resultant of noncoplanar force systems.

Resultants of Noncoplanar Force Systems

Force System	Resultant of System	Equations that Determine the Resultant
Concurrent (at point O)	A force, **R**, through point O.	$\mathbf{R} = \Sigma \mathbf{F}$ $(R_x = \Sigma F_x, R_y = \Sigma F_y, R_z = \Sigma F_z)$
Parallel (to z-axis)	A force, **R** (parallel to z-axis), or a couple, $\mathbf{C}^R$ ($\mathbf{C}^R$ has no z-component).	If $\Sigma F_z \neq 0$, the resultant is a force: $R = \Sigma F_z$ **R** acts through point $(\bar{x}, \bar{y})$ in the x-y plane: $\bar{x} = -\Sigma M_y / R$ and $\bar{y} = \Sigma M_x / R$
		If $\Sigma F_z = 0$, and $\Sigma M_O \neq 0$ (O is any point), the resultant is a couple: $\mathbf{C}^R = \Sigma \mathbf{M}_O$ $(C^R{}_x = \Sigma M_x, C^R{}_y = \Sigma M_y)$
General	A force **R**, a couple $\mathbf{C}^R$, or a wrench acting through point A.	If $\mathbf{R} \neq 0$ and $\Sigma M_O = 0$ (O is any point), the resultant is a force: $\mathbf{R} = \Sigma \mathbf{F}$ $(R_x = \Sigma F_x, R_y = \Sigma F_y, R_z = \Sigma F_z)$
		If $\mathbf{R} = \mathbf{0}$ and $\Sigma \mathbf{M}_O \neq 0$, the resultant is a couple: $\mathbf{C}^R = \Sigma \mathbf{M}_O (C^R{}_x = \Sigma M_x, C^R{}_y = \Sigma M_y, C^R{}_z = \Sigma M_z)$
		If $\mathbf{R} = \Sigma \mathbf{F}$, $\mathbf{C}^R = \Sigma \mathbf{M}_O$ (O is any point), and **R** and $\mathbf{C}^R$ are not perpendicular, the resultant can be reduced to a wrench. The steps to determine a wrench are as follows: 1. Determine $\mathbf{R} = \Sigma \mathbf{F}$ $(R_x = \Sigma F_x, R_y = \Sigma F_y, R_z = \Sigma F_z)$ 2. Determine $\mathbf{C}^R = \Sigma \mathbf{M}_O$ $(C^R{}_x = \Sigma M_x, C^R{}_y = \Sigma M_y, C^R{}_z = \Sigma M_z)$ 3. Resolve $\mathbf{C}^R$ into two components: $\mathbf{C}^R{}_t$ parallel to **R**: $\mathbf{C}^R{}_t = (\mathbf{C}^R \bullet \boldsymbol{\lambda})\, \boldsymbol{\lambda}$, where $\boldsymbol{\lambda}$ is a unit vector in the direction of **R** $\mathbf{C}^R{}_n$ normal ($\perp$) to **R**: $\mathbf{C}^R{}_n = \mathbf{C}^R - \mathbf{C}^R{}_t$ 4. Replace **R** and $\mathbf{C}^R{}_n$ by **R** acting through a point A. Locate A by $\mathbf{r} \times \mathbf{R} = \mathbf{C}^R{}_n$, where **r** is the position vector from point O to point A. 5. Show $\mathbf{C}^R{}_t$ at point A.

B. You Should be Able to:

- Determine the resultant of noncoplanar force systems.

C. Guided Practice Problems

S 11.1 Determine the resultant of the force system consisting of the three forces and one couple shown. Show the resultant on a sketch of the coordinate system.

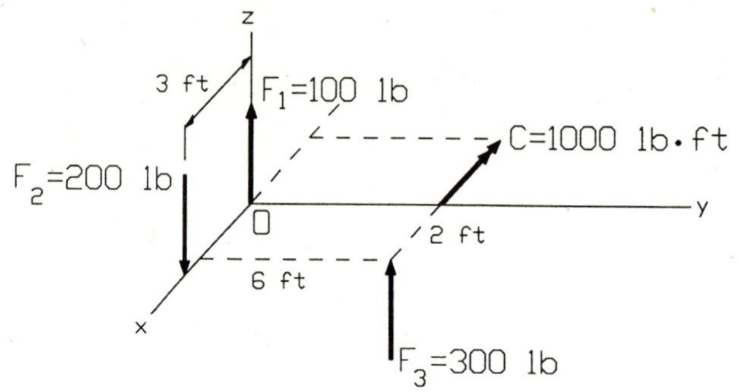

GUIDED SOLUTION

The force system is parallel, noncoplanar. Either scalar or vector equations can be used to solve this problem. Since the perpendicular distances to the coordinate axes are given, we will use scalar equations.

1. Sum the forces in the z-direction.

$+\uparrow \ R = \Sigma F_z =$

Since $\Sigma F_z \neq 0$, the resultant is a force.

2. Sum moments about the x- and y-axes. Note that there is no moment about the z-axis since all the forces are parallel to the z-axis.

$\Sigma M_x =$

$\Sigma M_y =$

The moments about the coordinate axes are the components of $\mathbf{C}^R = \Sigma \mathbf{M}_O$.

3. Determine the coordinates $\bar{x}$ and $\bar{y}$. (The coordinates of the point where R intersects the xy plane.)

$\bar{x} = -\Sigma M_y / R =$

$\bar{y} = \Sigma M_x / R =$

4. Show **R** on a sketch of the coordinate system.

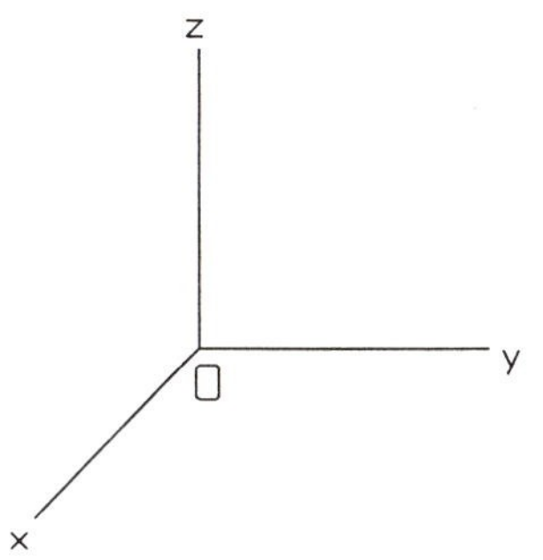

R = 200**k** lb
at $\bar{x} = 0$, $\bar{y} = 4.0$ ft

S 11.2 Determine the resultant of the force system consisting of the three forces shown. Write the resultant in vector form.

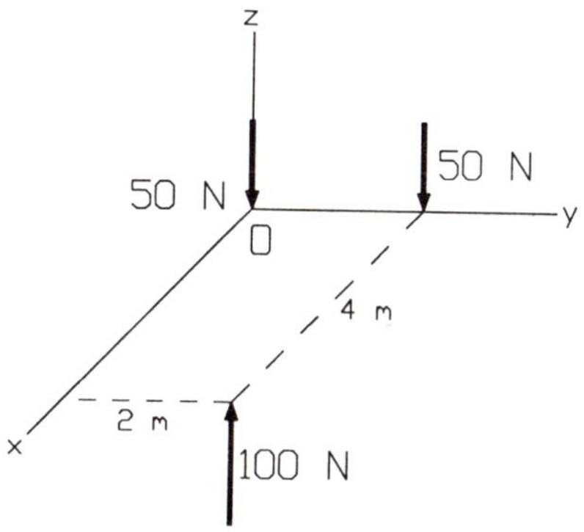

GUIDED SOLUTION

The force system is parallel, noncoplanar. Since the perpendicular distances to the coordinate axes are given, we will use scalar equations to solve this problem.

1. Sum the forces in the z-direction.

$+\uparrow \; R = \Sigma F_z =$

Since $\Sigma F_z = 0$, the resultant is not a force, but could be a couple.

2. Sum moments about the x- and y-axes. Note that there is no moment about the z-axis since all the forces are parallel to the z-axis.

$\Sigma M_x =$

$\Sigma M_y =$

3. Write the resultant in vector form.

$C^R =$

C^R = 100**i** – 400**j** N•m

S 11.3 A force-couple system consists of the force $\mathbf{R} = 300\mathbf{i} + 450\mathbf{j} + 200\mathbf{k}$ N, acting at the origin of a rectangular coordinate system, and the couple-vector $\mathbf{C}^R = 500\mathbf{i} + 400\mathbf{j} + 600\mathbf{k}$ N•m. Determine the equivalent wrench and the coordinates of the point where the axis of the wrench crosses the xy plane.

GUIDED SOLUTION

1. Determine the component of $\mathbf{C}^R$ along $\mathbf{R}$, $\mathbf{C}^R{}_t = (\mathbf{C}^R \bullet \boldsymbol{\lambda})\,\boldsymbol{\lambda}$.

(a) Write the unit vector in the direction of the axis of the wrench:

$\boldsymbol{\lambda} = \mathbf{R}/R =$

(b) Determine the magnitude of $\mathbf{C}^R{}_t$: $C^R{}_t = \mathbf{C}^R \bullet \boldsymbol{\lambda} =$

(c) Write the vector $\mathbf{C}^R{}_t$: $\mathbf{C}^R{}_t = C^R{}_t\,\boldsymbol{\lambda} =$

2. Determine the component of $\mathbf{C}^R$ that is perpendicular to $\mathbf{R}$, $\mathbf{C}^R{}_n$.

$\mathbf{C}^R{}_n = \mathbf{C}^R - \mathbf{C}^R{}_t =$

3. Locate the point where $\mathbf{R}$ intersects the xy plane. Let A be the point where $\mathbf{R}$ intersects the xy plane and let the position vector from the origin to point A(x,y) be $\mathbf{r}$.

(a) Complete the determinant $\mathbf{C}^R{}_n = \mathbf{r} \times \mathbf{R}$

$$\mathbf{C}^R{}_n = \begin{vmatrix} \mathbf{i} & \mathbf{j} & \mathbf{k} \\ x & y & 0 \\ & & \end{vmatrix}$$

(b) Expand the determinant.

(c) Equate the like components of the determinant with the components of $\mathbf{C}^R{}_n$ and solve for x and y.

(**i** component):

y =

(**j** component):

x =

$\mathbf{R} = 300\mathbf{i} + 450\mathbf{j} + 200\mathbf{k}$ N
$\mathbf{C}^R{}_t = 406\mathbf{i} + 609\mathbf{j} + 271\mathbf{k}$ N•m
x = 1.045 m and y = 0.470 m

S 12. Introduction to Distributed Normal Loads

Text Reference: Article 3.6; Sample Problems 3.11, 3.12

A. *You Should Understand:*

- This section deals with the special case of distributed loads that can be treated as parallel force systems.
- The resultant of a distributed force system acting on a load area is equal to the volume of the region between the load area and the load surface. The line of action of the resultant passes through the centroid of the volume of that region.
- The load intensity w of a line load is a function of the distance measured along the line of distribution.
- The resultant of a line load is the area under the load diagram and acts through the centroid of that area.
- If load areas or load diagrams have simple shapes, tables of areas and centroids can be used to determine the resultant forces and their lines of action.

B. *You Should be Able to:*

- Find the resultant of distributed loads.
- Locate the centroid of simple areas and shapes.
- Locate the centroid of an area and shape composed of a combination of simple areas and shapes.

C. Guided Practice Problems

S 12.1 Determine the resultant of the distributed load shown.

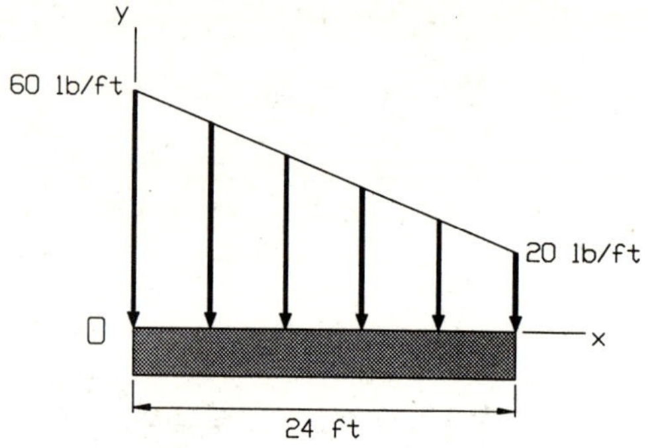

GUIDED SOLUTION

The load diagram can be represented by a rectangle (A_1) and a right-triangle (A_2), as shown below.

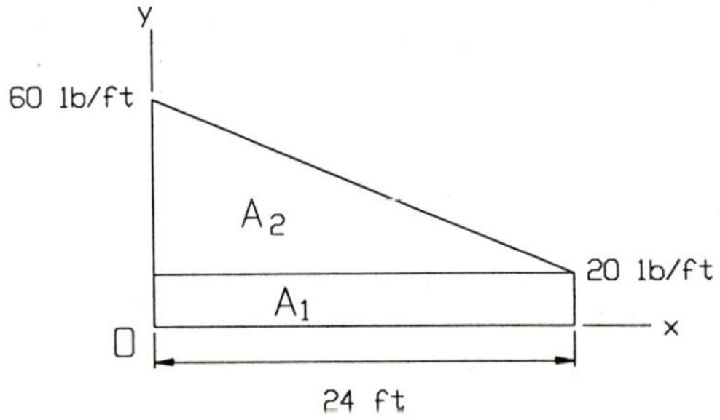

1. Calculate P_1 and P_2, the resultants of the line loads represented by A_1 and A_2, respectively.

✍ $P_1 = A_1 =$

$P_2 = A_2 =$

2. Determine the magnitude of the resultant load.

✍ $+\downarrow \; R = P_1 + P_2 =$

3. Determine the x-coordinates of the centroids of A_1 and A_2. (From Table 3.1 of the text)

✍ For A_1: $\bar{x}_1 =$

For A_2: $\bar{x}_2 =$

4. Show P_1 and P_2 on the sketch.

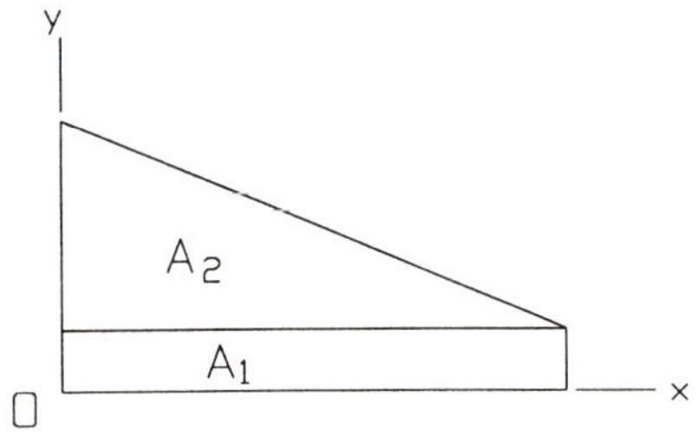

5. Determine $\bar{x}$, the x-ccordinate of the line of action of R.

✍ (a) Sum the moments of P_1 and P_2 about O:

$$\circlearrowleft + \quad \Sigma M_O = P_1 \bar{x}_1 + P_2 \bar{x}_2 =$$

✍ (b) Calculate $\bar{x}$: $\bar{x} = \Sigma M_O / R =$

R = 960 lb
$\bar{x}$= 10.0 ft

S 12.2 A triangular traffic sign experiences a uniformly distributed wind load of 720 N/m^2. Determine the resultant wind force exerted on the sign.

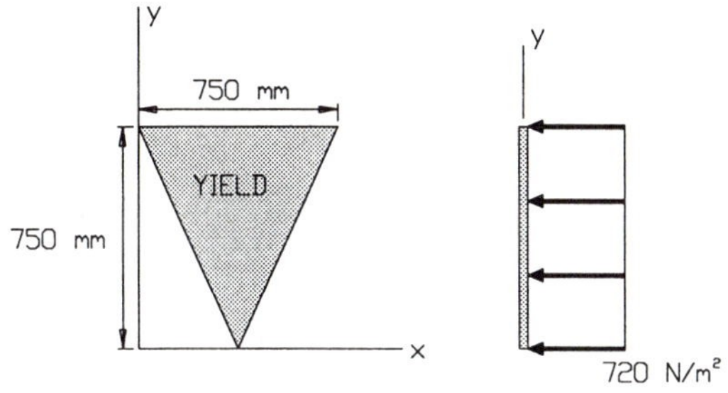

GUIDED SOLUTION

1. Determine the resultant wind force.

✍ $R = pA =$

2. Determine the coordinates ($\bar{x}$, $\bar{y}$) of the point through which R acts.

✍ By symmetry: $\bar{x}$=

From Table 3.1: $\bar{y}$=

R = 203 N
$\bar{x}$= 375 mm, $\bar{y}$= 500 mm

S 13. Free-Body Diagram of a Body

Text Reference: Article 4.3; Sample Problems 4.1 - 4.4

A. *You Should Understand:*

- A free-body diagram of a body is a sketch of the body showing all forces that act on the body.
- A reaction is the force exerted on the body by a support.
- An applied force is a force acting on the body (including weight) that is not exerted by a support.

B. *You Should be Able to:*

- Indicate the appropriate reactions applied by the various supports listed in Table 4.1 of the text.
- Draw a free-body diagram using the following steps:

    1. Sketch the body assuming that it has been disconnected from all its supports.
    2. Draw all the applied forces and label them. Include the weight of the body acting at the center of gravity of the body.
    3. Draw and label the reactions due to each of the disconnected supports. Assume the sense of a reaction force, if it is unknown.
    4. Indicate all relevant angles and dimensions on the sketch.

C. Guided Practice Problems

S 13.1 The homogeneous 20-kg bar AB is 1.5 m long. The bar is supported at A by a smooth pin and by a roller at B. Draw the free-body diagram of the bar and determine the number of unknowns.

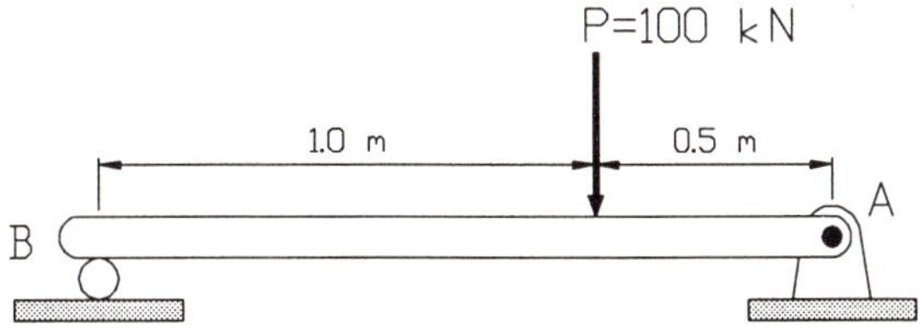

GUIDED SOLUTION

1. On the sketch below, draw the applied forces. Include the appropriate dimensions.

✍ Draw the weight acting at the center of the beam and the force P.

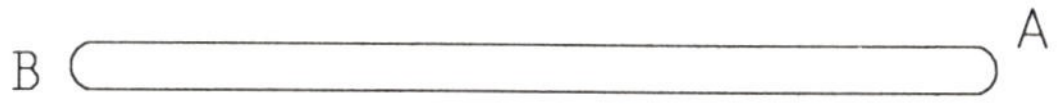

2. On the sketch above, draw the reactions.

✍ Draw the pin reaction at A (assume the direction of the vertical and horizontal components) and the roller reaction at B.

3. Identify the unknowns and count them.

✍ The unknowns are:

Number of unknowns =

Two components of pin reaction at A and vertical roller reaction at B. No. of unknowns = 3

S 13.2 The homogeneous bar BC weighs 100 lb and is 18 ft long. The bar is supported at C by a smooth pin and at B by a cable connected to the wall at A. The magnitude of P is 800 lb. Draw the free-body diagram of the bar and determine the number of unknowns.

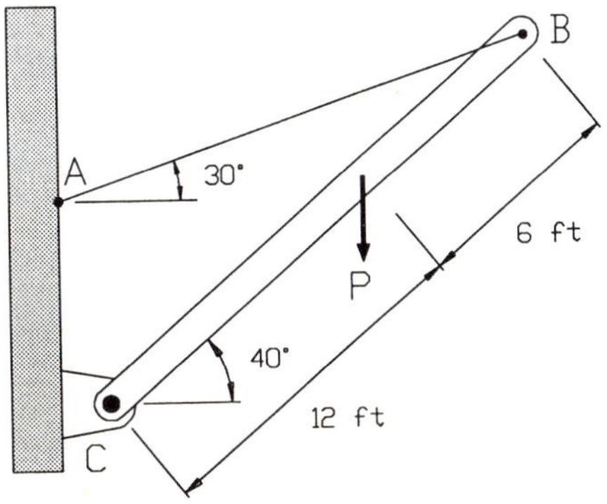

GUIDED SOLUTION

1. On the sketch below, draw the applied forces. Include the appropriate dimensions.

✍

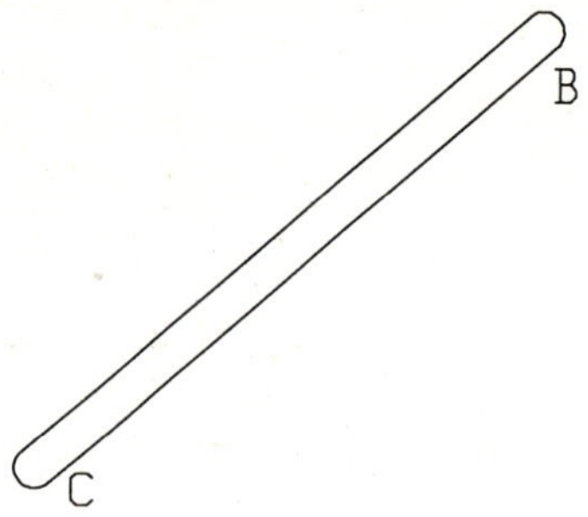

2. On the sketch above, draw the reactions.

✍ Draw the pin reaction at C and the cable tension at B.

3. Identify the unknowns and count them.

✍ The unknowns are:

Number of unknowns =

Two components of pin reaction at C and the cable tension at B. No. of unknowns = 3

S 14. WRITING AND SOLVING EQUILIBRIUM EQUATIONS FOR COPLANAR FORCE SYSTEMS

Text Reference: Articles 4.4, 4.5; Sample Problems 4.5 - 4.7

A. *You Should Understand:*

- A body is in equilibrium if the resultant of the force system acting on the body is zero.
- The number of equations required to determine the resultant of a force system is the same as the number of equations that guarantee that the resultant is zero.
- If the number of unknowns on a free-body diagram is the same as the number of available independent equilibrium equations, the force system is statically determinate.
- The following table reviews the equations that determine the resultant of coplanar force systems. It also provides examples of sets of independent equilibrium equations that ensure the resultant is zero. Note that the number of equations in each set matches the number of equations necessary to determine the resultant of the force system.

Resultants and Independent Equilibrium Equations for Coplanar Force Systems

Force System	Resultant of Force System	Equations that Determine the Resultant	Examples of Sets of Independent Equilibrium Equations	Number of Equations
Concurrent (through point O)	Force through point O	$R_x = \Sigma F_x$ $R_y = \Sigma F_y$	$\Sigma F_x = 0$ $\Sigma F_y = 0$	2
			$\Sigma M_A = 0$ $\Sigma M_B = 0$ (Line AB does not pass through point O)	
			$\Sigma F_{x'} = 0$ $\Sigma M_A = 0$ (A is any point except O; x' is not $\perp$ to line OA)	
Parallel (to y-axis)	Force or Couple	$R = \Sigma F_y$ $\Sigma M_O = Rd$	$\Sigma F_y = 0$ $\Sigma M_O = 0$ (O is any point)	2
			$\Sigma M_A = 0$ $\Sigma M_B = 0$ (line AB is not parallel to y-axis)	
General	Force or Couple	$R_x = \Sigma F_x$ $R_y = \Sigma F_y$ $\Sigma M_O = Rd$	$\Sigma F_x = 0$ $\Sigma F_y = 0$ $\Sigma M_O = 0$ (O is any point)	3
			$\Sigma M_A = 0$ $\Sigma M_B = 0$, $\Sigma F_{x'} = 0$ (x' is any direction except $\perp$ to line AB)	
			$\Sigma M_A = 0$ $\Sigma M_B = 0$, $\Sigma M_C = 0$ (A, B, C are not on the same line)	

B. You Should be Able to:

- Write independent equilibrium equations from a given free-body diagram and solve for the unknowns.

C. Guided Practice Problems

S 14.1 Figure (a) shows a weightless beam supported by a pin at A and a roller at B. The free-body diagram of the beam is shown in Figure (b). Find all the unknown forces acting on the beam.

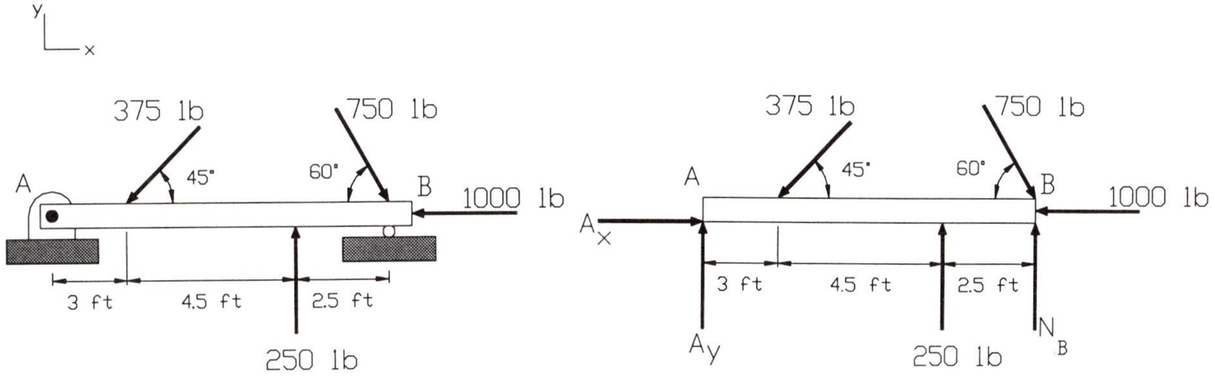

(a) Original

(b) Free-body Diagram

GUIDED SOLUTION

The force system acting on the beam is general coplanar. There are three independent equilibrium equations for such a system. There are three unknown forces acting on the beam, A_x, A_y, and N_B, as shown on the FBD. The system is therefore statically determinate, and all three unknowns can be found. In this problem, it is possible to write three independent equilibrium equations in such a way that each equation contains only one unknown.

1. Solve for A_x. Observe that A_x is the only unknown force acting in the x-direction. Therefore, summing forces in that direction will determine A_x.

$\overset{+}{\rightarrow}$ $\Sigma F_x = 0$:

2. Solve for N_B. Since N_B is the only unknown force that has a moment about point A, summing the moments about A will determine N_B.

$\circlearrowleft +$ $\Sigma M_A = 0$:

3. Solve for A_y. Since N_B has been determined, A_y is the only remaining unknown force acting in the y-direction. Therefore, summing forces in the y-direction will determine A_y.

$+\uparrow$ $\Sigma F_y = 0$:

4. Check your answer by summing moments about B. (The resultant should equal zero.)

$\circlearrowleft +$ $\Sigma M_B =$

$A_x = 890 \text{ lb} \rightarrow$, $N_B = 542 \text{ lb} \uparrow$, $A_y = 123 \text{ lb} \uparrow$

S 15. EQUILIBRIUM ANALYSIS FOR SINGLE-BODY PROBLEMS

Text Reference: Article 4.6; Sample Problems 4.8 - 4.10

A. *You Should Understand:*

- Equilibrium analysis for a single body combines the material covered in the previous two sections and consists of the following three steps:

  1. Drawing the free-body diagram of the body.
  2. Writing equilibrium equations.
  3. Solving the equilibrium equations for the unknowns.

B. *You Should be Able to:*

- Draw a free-body diagram of a single body.
- Using the FBD, write independent equilibrium equations.
- Solve the equilibrium equations for the unknowns.

C. Guided Practice Problems

S 15.1 The uniform rod ABC is 700 mm long and has a mass of 2.0 kg. The rod rests against smooth surfaces at A and B. Determine the magnitude of the reactions at A and B and the angle θ.

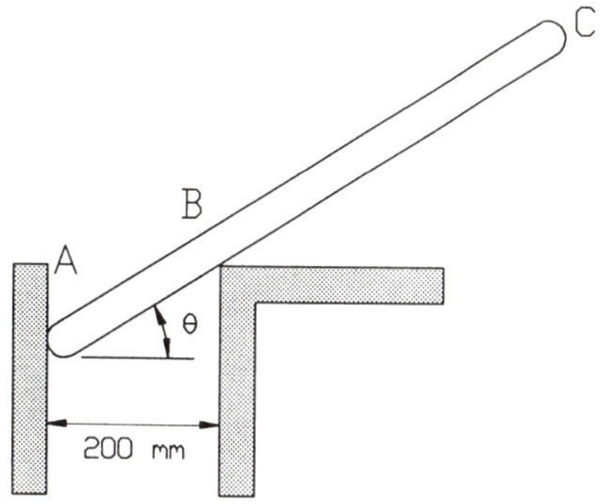

GUIDED SOLUTION

1. Draw the free-body diagram of the rod ABC. (Point G is the center of the rod.)

✍

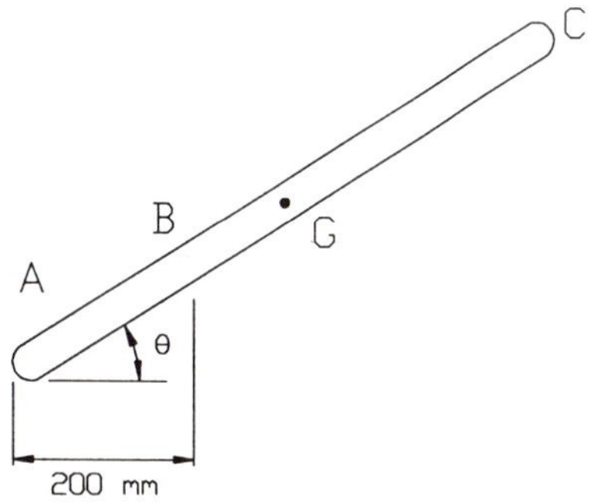

2. Write the three equilibrium equations. On the FBD, the unknowns are the normal forces at A and B, and the angle θ. Since the force system is general coplanar, there are three indpendent equilibrium equations. In this case, we choose a method of analysis that involves the following equations: $\Sigma F_y = 0$ and $\Sigma M_A = 0$ that include both the normal reaction at B and the angle θ, and $\Sigma F_x = 0$ that includes the normal reaction at A.

✍ (a) $+\uparrow$ $\Sigma F_y = 0$:

✍ (b) ↺+ $\Sigma M_A = 0$:

✍ (c) $\overset{+}{\rightarrow}$ $\Sigma F_x = 0$:

3. Solve the unknowns in the equilibrium equations.

✍ (a) Solve the equations from 2(a) and (b) for the normal force at B and angle θ.

✍ (b) Solve the equation from 2(c) for the normal force at A.

$N_B = 23.6$ N
$\theta = 33.9°$
$N_A = 13.2$ N

Another Method of Analysis

Another method that could have been used to solve for the three unknowns is as follows:

Equation	Unknowns	Solution
$\Sigma M_B = 0$	N_A and θ	Solve simultaneously for N_A and θ
$\Sigma F_{x'} = 0$ (x' is in the direction parallel to ABC)	N_A and θ	
$\Sigma F_y = 0$	N_B	Solve for N_B

S 16. Free-Body Diagrams Involving Internal Reactions

Text Reference: Article 4.7; Sample Problems 4.11, 4.12

A. *You Should Understand:*

- Internal reactions are forces that occur within a body. These reactions appear on the free-body diagram of the part of the body that has been isolated.
- When isolating two parts of a body so that an internal section is exposed, the reactions on each of the two exposed sections must be equal and opposite.
- Once the sense of an internal reaction has been assumed on a FBD, all subsequent FBDs must be consistent with the original assumption.
- Pin reactions are equal and opposite unless one of the following is true:
    1. External forces are applied to the pin.
    2. The weight of the pin is not negligible.
    3. More than two members are joined by the pin.
- The equilibrium equations for a whole system are not independent of the equilibrium equations for all separate parts of the system.

B. *You Should be Able to:*

- Draw the free-body diagram of a system as a whole.
- Draw the free-body diagram of any part of a system.
- Determine the number of independent equilibrium equations that are available for any or all parts of the system.

C. Guided Practice Problems

S 16.1 Draw the free-body diagram and determine the number of unknowns and independent equilibrium equations for (a) the entire frame, and (b) each of its parts.

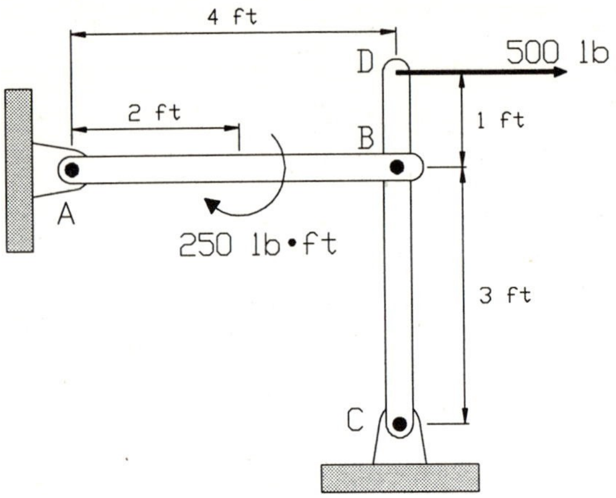

GUIDED SOLUTION

Note that weight is not given for any part of the frame. The weight of each part is therefore assumed negligible. In other words, each part is treated as 'weightless'.

Part (a)

1. <u>Draw the free-body diagram of the entire frame.</u>

✍

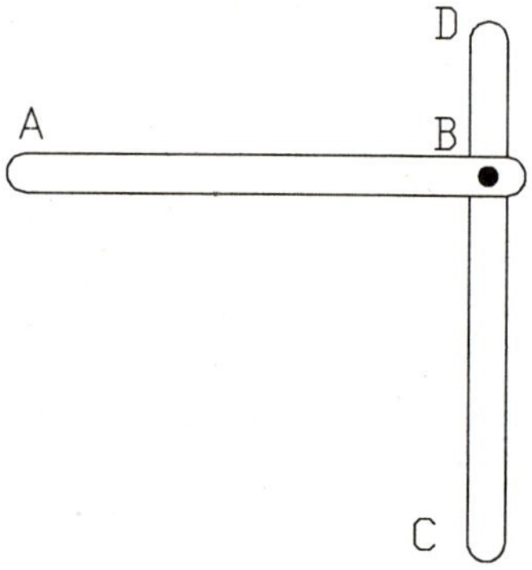

2. <u>Determine the number of independent equilibrium equations and the number of unknowns on the FBD.</u> This is a general coplanar force system.

✍ (a) Number of independent equilibrium equations =

✍ (b) The number of unknowns on the FBD =

3 equations and 4 unknowns

Part (b)

1. Draw the free-body diagram of each part of the frame.

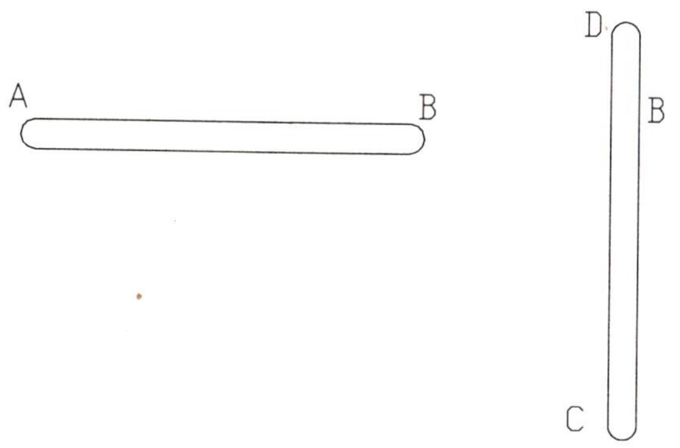

2. Determine the number of independent equilibrium equations and the total number of unknowns on the FBD's. The force system for each part of the frame is general coplanar.

(a) Number of independent equilibrium equations =

(b) The number of unknowns on the FBDs =

6 equations and 6 unknowns

S 16.2 The 20-kg block D is smooth and bar AB has a mass of 10 kg. A cable connects end B of the bar with the block. Draw the free-body diagram and determine the number of independent equilibrium equations and unknowns for (a) the entire system (with the pulleys removed), and (b) each of its parts.

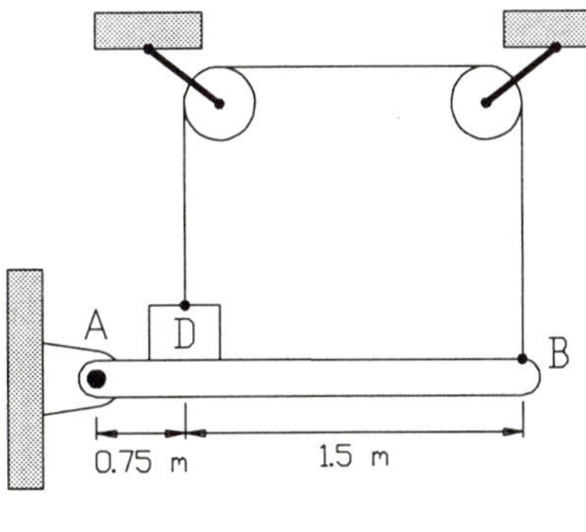

GUIDED SOLUTION

Part (a)

1. Draw the free-body diagram of the entire system (with the pulleys removed). Recall that a pulley simply changes the direction of a cable, but does not affect the magnitude of the tension in the cable.

✍

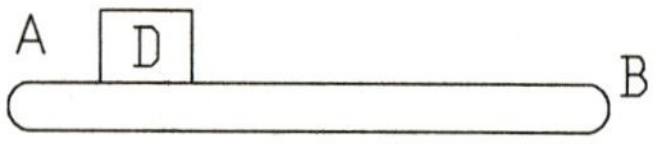

2. Determine the number of independent equilibrium equations and the number of unknowns on the FBD. This is a general coplanar force system.

✍ (a) Number of independent equilibrium equations =

✍ (b) The number of unknowns on the FBD =

3 equations and 3 unknowns

Part (b)

1. Draw the free-body diagram of each part of the system.

✍

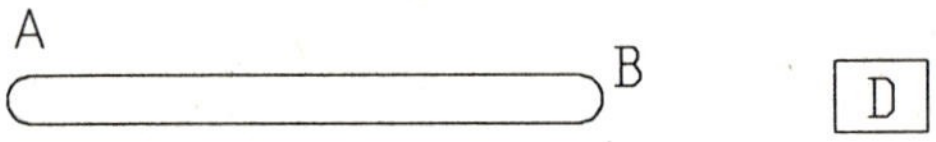

2. Determine the number of independent equilibrium equations and the total number of unknowns on the FBD's. Bar AB forms a general coplanar force system and block D is a particle with collinear forces.

✍ (a) Number of independent equilibrium equations =

✍ (b) The number of unknowns on the FBDs =

4 equations and 4 unknowns

S 17. EQUILIBRIUM ANALYSIS OF COMPOSITE BODIES

Text Reference: Article 4.8; Sample Problems 4.13, 4.14

A. You Should Understand:

- Analysis of composite bodies often involves considering more than one free-body diagram.
- Equilibrium analysis of a composite body should usually begin with the free-body diagram of the whole body, with external reactions calculated as needed. The analysis then proceeds to one or more parts of the body.
- Equilibrium equations should be written after you determine that the total number of available equilibrium equations, based on the type of force systems represented by the free-body diagram(s), matches the total number of unknowns.

B. You Should be Able to:

- Perform an equilibrium analysis of a composite body and its parts.

C. Guided Practice Problems

S 17.1 Determine the magnitude and direction of the pin reaction at C in problem S 16.1.

GUIDED SOLUTION

The usual first step in the analysis of a composite body is to draw the free-body diagram of the whole body.

1. Draw the free-body diagram of the entire frame.

✍

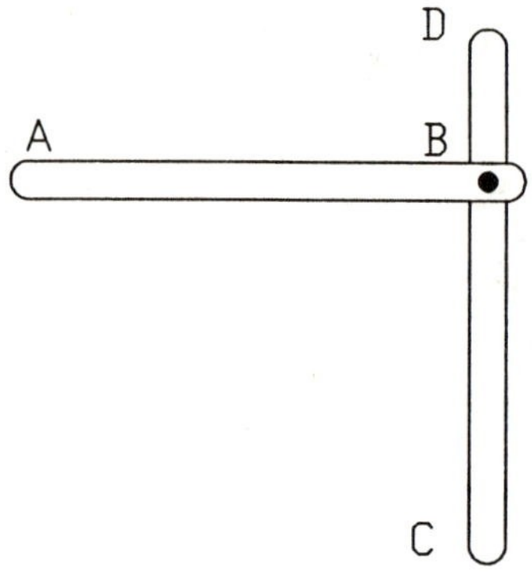

The FBD of the whole frame contains four unknowns. The number of equilibrium equations for a general force system is three. Since these two numbers do not match, we should draw another FBD before writing any equations. Since we are asked to find the pin reaction at C, we choose to draw the free-body diagram of bar CD as the next step.

2. Draw the free-body diagram of bar CD.

✍

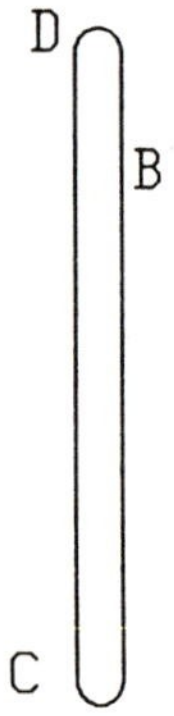

3. For both the entire frame and bar CD, determine and compare the total number of equilibrium equations and the total number of unknowns on the FBDs.

✍ (a) The total number of independent equilibrium equations =

✍ (b) The number of unknowns on the FBDs =

The next step in the equilibrium analysis is to study the two FBDs and decide which equilibrium equations to use. Notice that from the FBD of the whole body, the equation $\Sigma M_A = 0$ will contain the components of the pin reaction at C. The equation $\Sigma M_B = 0$ from the FBD of bar CD will also contain the components of the pin reaction at C. These two equations can then be solved simultaneously for the pin reaction components.

The mathematical details follow.

4. Write the independent equilibrium equations required to solve for C_y and C_x and solve.

✍ (a) From the FBD of the entire frame:

↶+ $\Sigma M_A = 0$: (1)

✍ (b) From the FBD of bar CD:

↶+ $\Sigma M_B = 0$: (2)

✍ (c) Solve Eqn. (1) and Eqn. (2) simultaneously.

5. Determine the magnitude and direction of the resultant pin reaction on the bar at C.

✍ Magnitude of the reaction, R_C:

$$R_C = \sqrt{C_x^2 + C_y^2} =$$

The angle θ, counterclockwise from the horizontal:

$$\theta = \tan^{-1}(C_y / C_x) =$$

$R_C = 178$ lb, $\theta = 20.6°$

Another Method of Analysis

We could have drawn the FBD of bar AB instead of bar CD.

S 17.2 Determine the magnitude of the reaction between block D and bar AB in problem S 16.2.

GUIDED SOLUTION

As before, the first step in any method of analysis of a composite body should be to draw the free-body diagram of the whole body.

1. Draw the free-body diagram of the entire body.

Notice that the number of unknowns on the FBD of the whole body is three. The number of independent equilibrium equations is also three. Since these two numbers match, it is possible to solve for all the unknowns: the components of the pin reaction at A, as well as the tension in the cable. To find the reaction between D and bar AB, however, we need a FBD that shows this force. A FBD of either block D or bar AB will show the required force. We choose to draw the FBD of block D.

2. Draw the free-body diagram of block D.

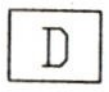

3. Write the independent equilibrium equations required to solve for N_D and solve.

(a) From the FBD of the entire system, solve for the cable tension, T.

↻+ $\Sigma M_A = 0$:

(b) From the FBD of block D, solve for N_D.

+↑ $\Sigma F_y = 0$:

$N_D = 110$ N

Another Method of Analysis

The FBD of bar AB could have been drawn instead of block D.

S 17.3 Neglecting the weights of the bars and pulley, determine the magnitude of the pin reaction at B. All the dimensions are in millimeters. Bars AC and BD are pinned at E and the cable connecting the 150-kg load with end C is wound around the pulley. Another cable connects the center of the pulley, point O, with end D.

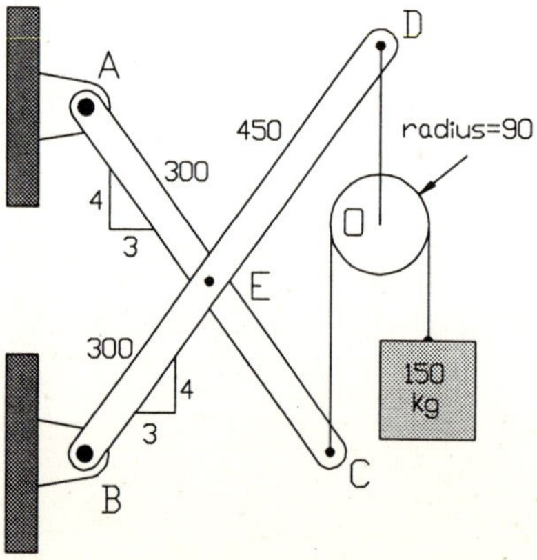

GUIDED SOLUTION

1. Draw the FBD of the entire frame.

✍

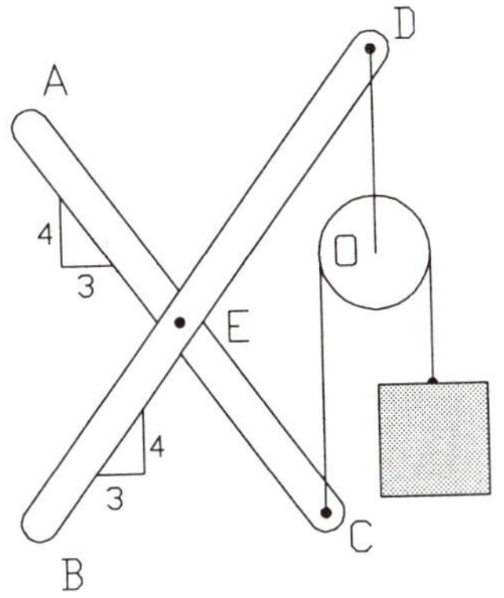

2. Count the unknowns on the FBD of the entire frame and compare the number of unknowns to the number of equilibrium equations.

✍ The number of unknowns on the FBD =

Number of independent equilibrium equations =

Since there are more unknowns on the FBD than independent equilibrium equations, we will need to draw another FBD. However, we note that the equation $\Sigma M_A = 0$ (Eqn. 1) gives the horizontal component of the pin reaction at B (B_X). We need the pin reaction at B and so for the next FBD, we choose the part of the frame that includes everything except bar AC.

3. Draw the FBD of BD, the pulley and the load.

✍

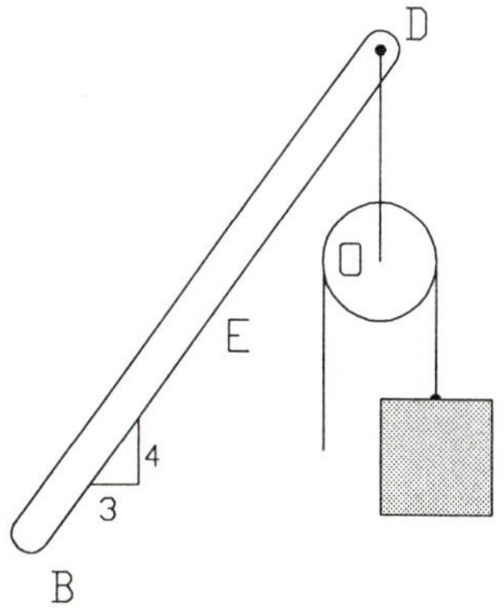

4. Count the unknowns on both free-body diagrams and compare the number of unknowns to the number of equilibrium equations.

✍ The number of unknowns on the FBDs =

Number of independent equilibrium equations =

Once again, there are more unknowns on both FBDs than independent equilibrium equations. Therefore, we need another FBD. We note that if we know B_X (from Eqn. 1), the equation $\Sigma M_E = 0$ (Eqn. 2) gives two unknowns, the vertical pin reaction component at B (B_y) and the tension (T_1) in the cable supporting the 150-kg load. It appears that drawing another FBD that contains that cable tension will help. We draw the FBD of the pulley.

5. Draw the FBD of the pulley.

✍

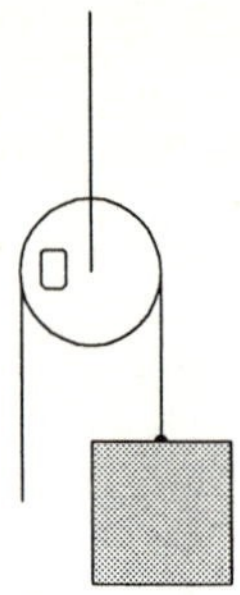

6. Count the unknowns on all three free-body diagrams and compare the number of unknowns to the number of equilibrium equations.

✍ The number of unknowns on the FBDs =

Number of independent equilibrium equations =

The total number of unknowns on all the FBDs matches the number of independent equilibrium equations. Note that the equation $\Sigma M_O = 0$ (Eqn. 3) gives the tension (T_2) in the cable connecting O and D.

The mathematical details follow.

7. Write the idependent equilibrium equations required to solve for B_x and B_y and solve.

✍ Using Eqn. 1, from the FBD of the whole frame, solve for B_x.

+ $\Sigma M_A = 0$:

Using Eqn. 3, from the FBD of the pulley, solve for T_2.

+ $\Sigma M_O = 0$:

Using Eqn. 2, from the FBD of the frame part, solve for B_y.

+ $\Sigma M_E = 0$:

8. Determine the magnitude of the resultant pin reaction at B, R_B.

$R_B = \sqrt{B_x^2 + B_y^2} =$

$R_B = 2760$ N

Other Methods of Analysis

In addition to the FBD of the entire frame, this problem requires the FBDs of two other parts of the system. The problem could have been solved with the same number of steps if the second FBD had been just bar BD or if it had been just bar AC.

S 18. TWO-FORCE AND THREE-FORCE BODIES

Text Reference: Article 4.9; Sample Problems 4.15, 4.16

A. You Should Understand:

- Two forces that hold a body in equilibrium must be equal in magnitude and oppositely directed along the same line of action.
- A body held in equilibrium by only two forces is called a two-force body.
- The use of the Two-force Principle in equilibrium analysis is convenient, but not necessary. The primary advantage in recognizing two-force bodies is that the required number of equilibrium equations is reduced.
- Three non-parallel, coplanar forces that hold a body in equilibrium must be concurrent.
- A body held in equilibrium by only three non-parallel, coplanar forces is called a three-force body.
- The use of the Three-force Principle is inconvenient when it is difficult to locate the point of concurrency.

B. You Should be Able to:

- Recognize two-force bodies.
- Use two-force bodies in equilibrium analysis.
- Recognize three-force bodies.
- Use three-force bodies in equilibrium analysis.

C. Guided Practice Problems

S 18.1 Determine the pin reaction at B. Neglect the weight of the bars and pulley.

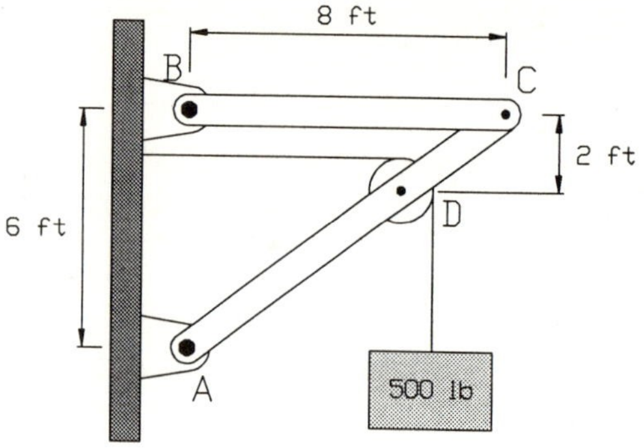

GUIDED SOLUTION

The recommended first step in the analysis of a composite body is to draw the free-body diagram of the entire body. In this case, we must disconnect (or cut) the cable and show the tension in the cable. (Note that since a pulley only changes direction, the force in the cable is 500 lb.)

1. Draw the free-body diagram of the entire body (frame) below. When you draw the FBD, note that bar BC is acted upon by only two forces: the resultant pin reaction at B and the resultant pin reaction at C. Therefore, bar BC is a two-force body. Label the resultant pin reaction at B as F_{BC}. (F_{BC} will lie along the line BC.)

✍

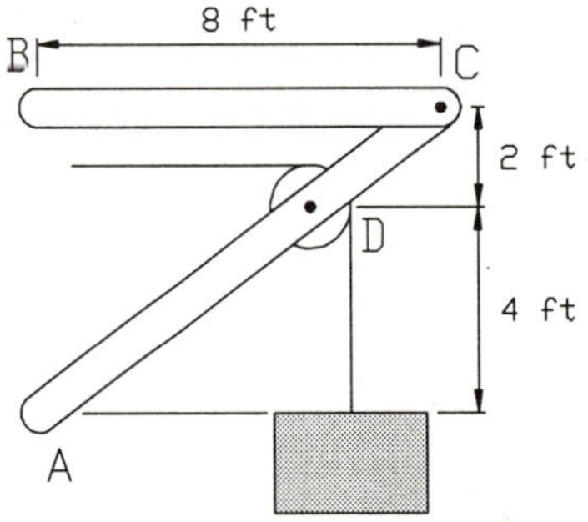

2. Compare the total number of unknowns with the number of equilibrium equations.

✍ (a) The number of unknowns on the FBD of the entire frame =

✍ (b) Total number of independent equilibrium equations =

3. Write and solve an equilibrium equation that will yield F_{BC}, the pin reaction at B.

✍ ↺+ $\Sigma M_A = 0$:

$F_{BC} = 111$ lb ←

S 18.2 Determine the reactions at A and C. Neglect the weight of the member.

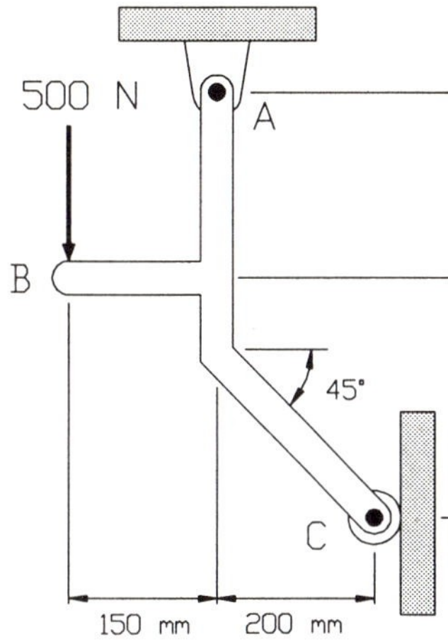

Member ABC is a three-force body. The resultant pin reaction at A, the normal force at C, and the applied force at B intersect at a point O. Note that since the line of the resultant pin reaction passes through point O, the direction of the line is known.

1. Draw the free-body diagram of the entire body below. Show the orientation of the resultant pin reaction at A and the point of concurrency, point O.

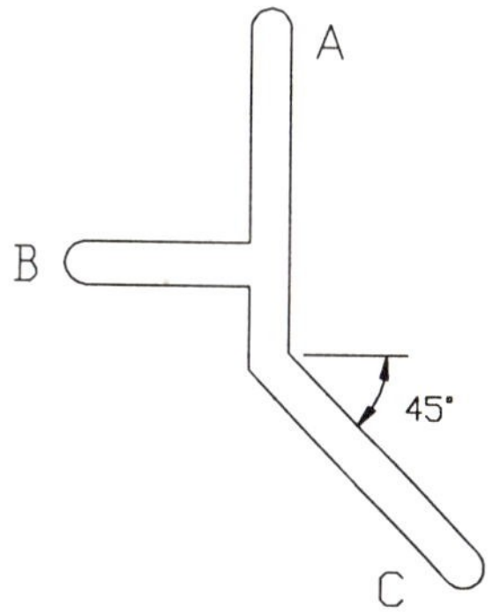

2. Compare the total number of unknowns with the number of equilibrium equations.

✍ (a) The number of unknowns on the FBD of the entire frame =

✍ (b) Total number of independent equilibrium equations =

3. Write and solve the equilibrium equations.

✍ (a) Solve the equation that will yield the resultant pin reaction at A.

$+\uparrow \ \Sigma F_y = 0$:

✍ (b) Solve the equation that will yield the reaction at C.

$\xrightarrow{+} \ \Sigma F_x = 0$:

$R_A = 527$ N
$R_C = 167$ N $\leftarrow$

S 19. TRUSSES, METHOD OF JOINTS

Text Reference: Articles 4.10, 4.11; Sample Problem 4.17

A. *You Should Understand:*

- The analysis of trusses is based on the following three assumptions:
    1. The weights of the members of a truss are negligible.
    2. All joints in a truss are smooth pins.
    3. All forces applied to a truss act at the joints.

- Applying the assumptions, each member of a truss is a two-force body.

- A truss member is said to be in tension if the force acting on it tends to elongate, or stretch, it. A truss member is in compression if the force acting on it tends to compress, or shorten, it.

- The method of joints refers to the equilibrium analysis of individual joints of a truss.

- The forces shown on a FBD of a joint of a plane truss are concurrent, coplanar. Therefore, there are two independent equilibrium equations for the FBD of a joint.

B. *You Should be Able to:*

- Draw a free-body diagram of a joint of a truss.

- Use the method of joints to determine the forces in members of a truss.

C. Guided Practice Problems

S 19.1 For the plane truss shown below, use the method of joints to determine the force in members AB, AH, BH, and BC. The truss is supported by a pin at A and a roller at E.

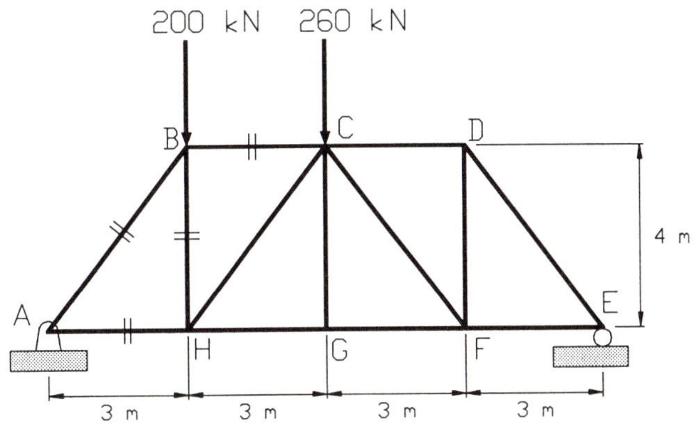

GUIDED SOLUTION

A good place to begin the analysis of a truss is with the free-body diagram of the entire truss.

1. Draw the free-body diagram of the entire truss below.

✍

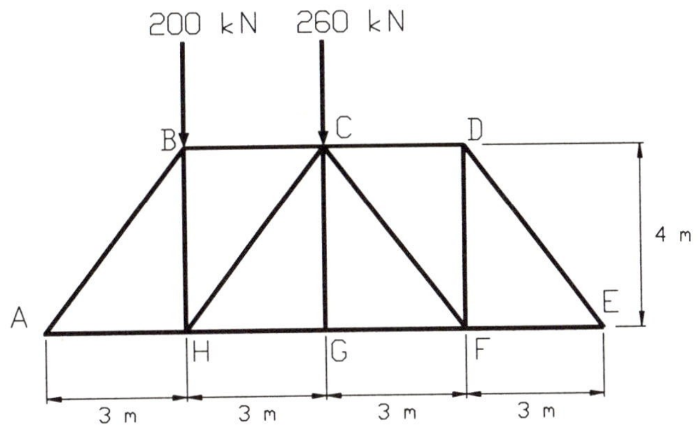

2. Count the number of unknowns on the FBD and the number of equilibrium equations.

✍ The number of unknowns on the FBD of the entire truss =

Number of independent equilibrium equations =

Can the external reactions be determined from this FBD?

3. From the FBD of the entire truss, determine the components of the pin reaction at A.

✍ (a) Solve for the vertical component.

↺+ $\Sigma M_E = 0$:

✍ (b) Solve for the horizontal component.

$\xrightarrow{+}$ $\Sigma F_x = 0$:

4. Draw the free-body diagram of joint A below. We draw joint A because it exposes the force in AB and in AH. Assume each of the members is in tension.

✍

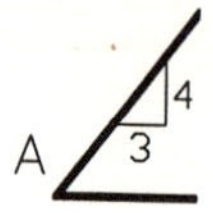

5. From the FBD of joint A, determine the force in members AB and AH. (Indicate whether the member is in tension or compression)

✍ (a) Determine the force in member AB.

$+\uparrow \qquad \Sigma F_y = 0:$

✍ (b) Determine the force in member AH.

$\overset{+}{\longrightarrow} \qquad \Sigma F_x = 0:$

6. Draw the free body diagram of joint B below. We draw joint B because it exposes the force in BH and BC. Recall that we have already solved for the force in member AB (part 5, above).

✍

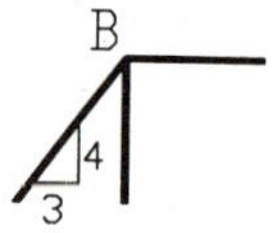

7. From the FBD of joint B, determine the force in members BH and BC. (Indicate whether the member is in tension or compression)

✍ (a) Solve for the force in member BH.

$+\uparrow \qquad \Sigma F_y = 0:$

✍ (b) Solve for the force in member BC.

$\overset{+}{\longrightarrow} \qquad \Sigma F_x = 0:$

AB = – 350 kN, AH = 210 kN, BH = 80 kN, BC = – 210 kN

Note: A positive sign for a force in a two-force body indicates that the member is in tension, and a negative sign indicates that the member is in compression. If each exposed member is assumed to be in tension when drawing the FBD of a joint, the resulting sign for the force in the member will automatically follow this convention.

S 20. METHOD OF SECTIONS

Text Reference: Article 4.12; Sample Problem 4.18

A. You Should Understand:

- Equilibrium analysis of a part of a truss that contains more than one joint (i.e. a section) is called the Method of Sections.
- The free-body diagram of a plane truss section is a nonconcurrent, coplanar force system. Therefore, there are three independent equilibrium equations.
- When applying the Method of Sections, a cut is made through members to completely isolate a section of the truss. The free-body diagram shows all of the forces acting on the section, including the forces in the exposed members.
- It is convenient, if possible, to isolate sections in such a way that no more than three members are exposed. In that way, the forces in the three exposed members can be determined from the one FBD (assuming external reactions have been previously calculated).

B. You Should be Able to:

- Draw free-body diagrams of truss sections.
- Write and solve the equilibrium equations to determine the force in any member of a truss.

C. *Guided Practice Problems*

S 20.1 Using the Method of Sections, determine the force in members BH and BC of the truss in problem S 19.1.

GUIDED SOLUTION

A good place to start the analysis is with the free-body diagram of the entire truss. The determination of the components of the external pin reaction at A were already done in S19.1 and will not be repeated here (the reaction at E was determined from the equation $\Sigma F_y = 0$).

1. Decide where to make the cut to expose the force in members BH and BC. If possible, we want to expose the force in BH and in BC without cutting through more than three members. The cut 1-1 meets this criterion.

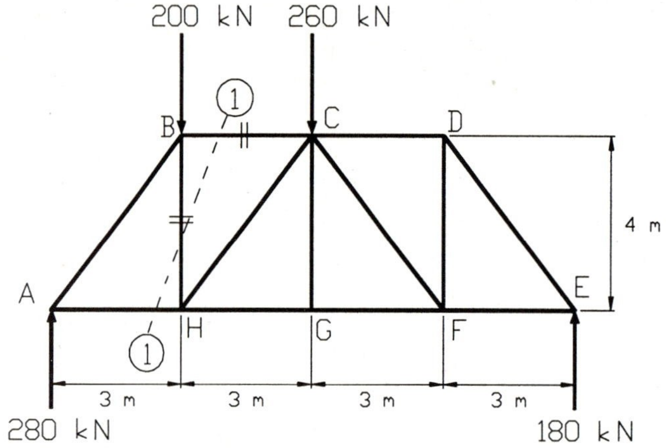

2. Draw a free-body diagram of a section isolated by cut 1-1. We can draw the FBD of the section to the left or right of cut 1-1. We choose to draw the FBD of the section to the left of cut 1-1. Assume the forces are tensile.

✍

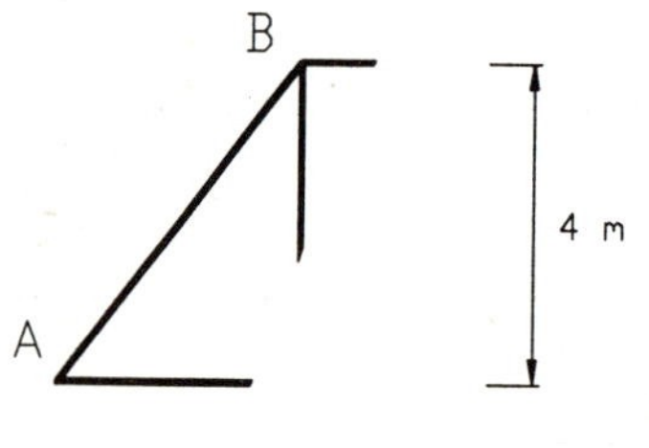

3. Write the equilibrium equations to solve for the force in members BH and BC. Note that the FBD of the section contains three unknowns. Therefore, all three unknowns can be determined.

✍ (a) Solve for the force in member BH.

$+\uparrow \qquad \Sigma F_y = 0:$

✍ (b) Solve for the force in member BC.

$\circlearrowleft + \qquad \Sigma M_A = 0:$

BH = 80 kN, BC = – 210 kN

S 21. NONCOPLANAR EQUILIBRIUM (FREE-BODY DIAGRAMS)

Text Reference: Articles 5.2, 5.3; Sample Problems 5.1 - 5.3

A. You Should Understand:

- The use of vector notation is frequently advantageous for noncoplanar equilibrium analysis.
- Six scalar equations are required to determine the resultant of a general, noncoplanar force system. Therefore, six equilibrium equations are required to guarantee that the resultant is zero.
- The procedure for constructing a free-body diagram of a body, or connected bodies, acted upon by a noncoplanar force system is identical to that used for a body acted upon by a coplanar force system.
- Table 5.1 of the text lists the reactions for noncoplanar supports.

B. You Should be Able to:

- Draw free-body diagrams of bodies (single and composite) that are acted upon by noncoplanar force systems.

C. Guided Practice Problems

S 21.1 A mast of negligible weight is supported by a ball-and-socket joint at A and two cables, as shown. Draw the free-body diagram of the mast and identify the unknowns.

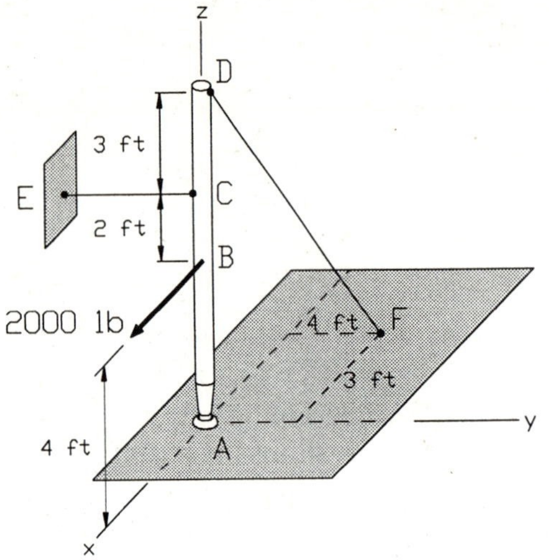

GUIDED SOLUTION

1. On the sketch below, draw the FBD of the mast.

✍ Show the following applied forces: 2000-lb force
T_{DF}, the tension in DF
T_{CE}, the tension in CE

Show the reaction at the ball-and-socket joint at A: A_x, A_y, A_z

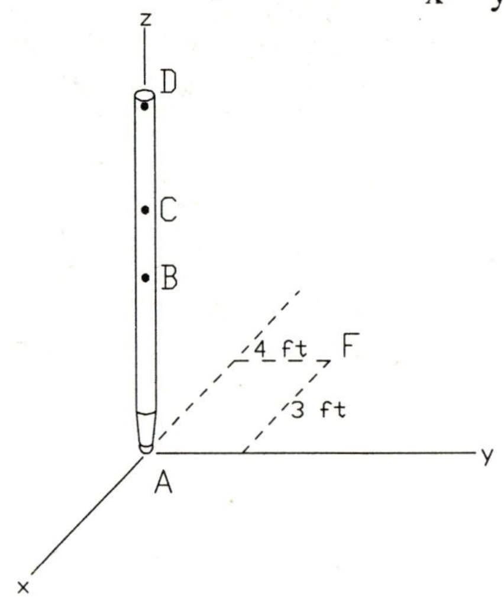

2. Identify and count the unknowns.

✍ (a) The unknowns are:

✍ (b) The number of unknowns =

Three components of the ball-and-socket joint: A_x, A_y, A_z, and two cable tensions: T_{DF}, T_{CE}.
No. of unknowns = 5

S 22. Writing and Solving Equilibrium Equations for Noncoplanar Force Systems

Text Reference: Articles 5.4 - 5.6; Sample Problems 5.4 - 5.7

A. *You Should Understand:*

- The number of equations required to determine the resultant of a force system is the same as the number of equations that guarantee the resultant is zero.
- One set of equilibrium equations for a general, noncoplanar force system is:

$$\Sigma F_x = 0, \quad \Sigma F_y = 0, \quad \Sigma F_z = 0,$$
$$\Sigma M_x = 0, \quad \Sigma M_y = 0, \quad \Sigma M_z = 0.$$

- Writing a moment equation about an axis is frequently convenient for noncoplanar force systems because forces that intersect the axis do not appear in the equation.
- The following table reviews the resultants of different types of noncoplanar force systems. The number of independent equilibrium equations for each type is also given.

Resultants and Number of Independent Equilibrium Equations for Noncoplanar Force Systems

Noncoplanar Force System	Resultant	Number of Independent Equilibrium Equations
General	Wrench (force and couple)	6
Concurrent at point O	Force **R** through point O	3
Parallel (to z-axis)	Force parallel to z-axis, or couple-vector perpendicular to z-axis	3
All forces intersect an axis	Force or Couple (no component along the axis)	5

B. *You Should be Able to:*

- Write independent equilibrium equations for a free-body diagram that contains a noncoplanar force system, and solve for the unknowns.

C. Guided Practice Problems

S 22.1 The free body diagram of the mast from problem S 21.1 is shown below. Determine the magnitude of the reaction at the ball-and-socket joint at A and the tensions in the two cables, T_{CE} and T_{DF}.

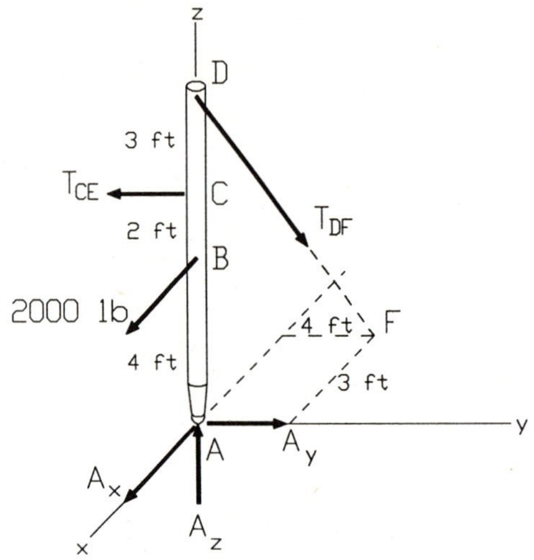

GUIDED SOLUTION

All the forces intersect the z-axis. Therefore, there are five independent equilibrium equations. Since there are five unknowns on the FBD (A_x, A_y, A_z, T_{CE} and T_{DF}), the system is statically determinate, and all five unknowns can be found.

For the method of analysis, we study the FBD for the most convenient equations to write. We choose to solve $\Sigma \mathbf{M}_A = \mathbf{0}$ for T_{CE} and T_{DF} because the two components (**i** and **j**) of the equation will enable us to solve for the unknowns. Recall that since all the forces intersect the z-axis, the moment equation has no component in the z direction. The equation $\Sigma \mathbf{F} = \mathbf{0}$ will be used to solve for the components of the reaction at A.

The mathematical details follow.

1. <u>Write $\Sigma \mathbf{M}_A = \mathbf{0}$ and solve for T_{CE} and T_{DF}.</u> $(\Sigma \mathbf{M}_A = \Sigma(\mathbf{r} \times \mathbf{F}) = \mathbf{0})$

✍ (a) Write the 2000-lb force in vector form: $\mathbf{F} =$

✍ (b) Write $\mathbf{T}_{CE}$: $\mathbf{T}_{CE} = T_{CE}\,\boldsymbol{\lambda}_{CE} =$

✍ (c) Write $\mathbf{T}_{DF}$: $\mathbf{T}_{DF} = T_{DF}\,\boldsymbol{\lambda}_{DF} =$

✍ (d) Choose $\mathbf{r}$ for the 2000-lb force and write it vector form: $\mathbf{r}_{AB} =$

✍ (e) Choose $\mathbf{r}$ for $\mathbf{T}_{CE}$ and write it in vector form: $\mathbf{r}_{AC} =$

✍ (f) Choose $\mathbf{r}$ for $\mathbf{T}_{DF}$ and write it in vector form: $\mathbf{r}_{AD} =$

✍ (g) Write $\Sigma \mathbf{M}_A = \mathbf{0}$: $\Sigma(\mathbf{r} \times \mathbf{F}) = \mathbf{0}$

$$\begin{vmatrix} \mathbf{i} & \mathbf{j} & \mathbf{k} \\ & & \\ & & \end{vmatrix} + \begin{vmatrix} \mathbf{i} & \mathbf{j} & \mathbf{k} \\ & & \\ & & \end{vmatrix} + \begin{vmatrix} \mathbf{i} & \mathbf{j} & \mathbf{k} \\ & & \\ & & \end{vmatrix} = \mathbf{0}$$

✍ (h) Equate the like components:

$\Sigma(\mathbf{i}$ components$) = 0$:

$\Sigma(\mathbf{j}$ components$) = 0$:

Solve for T_{CE} and T_{DF}:

2. Solve for the resultant reaction at A. The components of the equation $\Sigma \mathbf{F} = \mathbf{0}$ will each solve for one unknown.

✍ (a) Calculate A_x: $\Sigma F_x = 0$:

✍ (b) Calculate A_y: $\Sigma F_y = 0$:

✍ (c) Calculate A_z: $\Sigma F_z = 0$:

✍ (d) Calculate the resultant force at A: $A = \sqrt{A_x^2 + A_y^2 + A_z^2} =$

$T_{CE} = 1780$ lb
$T_{DF} = 3050$ lb
$A = 2950$ lb

Another Method of Analysis

Another method of analysis is: $\Sigma M_{AF} = 0$ to determine T_{CE}, $\Sigma M_y = 0$ to solve for T_{DF}.

S 23. EQUILIBRIUM ANALYSIS

Text Reference: Article 5.7; Sample Problems 5.8 -5.10

A. *You Should Understand:*

- Equilibrium analysis of bodies subjected to noncoplanar force systems is a combination of the topics discussed in the previous two sections. The steps involved in the analysis are:

  1. Draw the free-body diagrams of single or composite bodies.

  2. Write equilibrium equations.

  3. Solve the equilibrium equations for the unknowns.

B. *You Should be Able to:*

- Solve equilibrium problems of bodies, or parts of bodies, subjected to noncoplanar force systems.

C. Guided Practice Problems

S 23.1 The boom OA is 15 ft long, has negligible weight and lies in the y-z plane. It is supported by a ball-and-socket joint at O and two cables connected to the wall at B and C. Determine the magnitude of the tensions in the two cables and the resultant reaction at O.

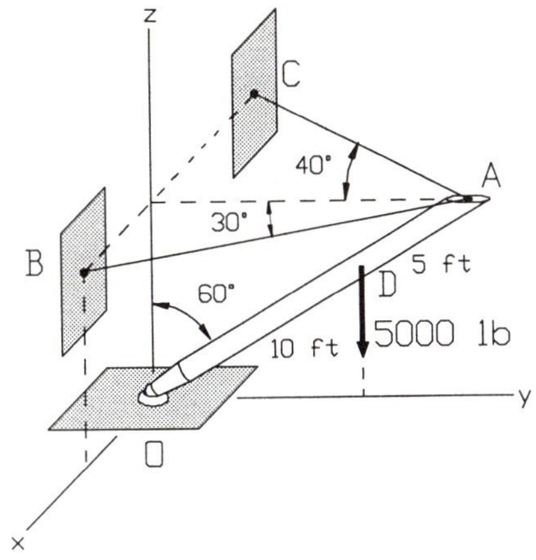

GUIDED SOLUTION

1. <u>Draw the free-body diagram of boom OA</u>. (Label the tensions T_{AC} and T_{AB} and the components of the pin reaction O_x, O_y, O_z)

✍

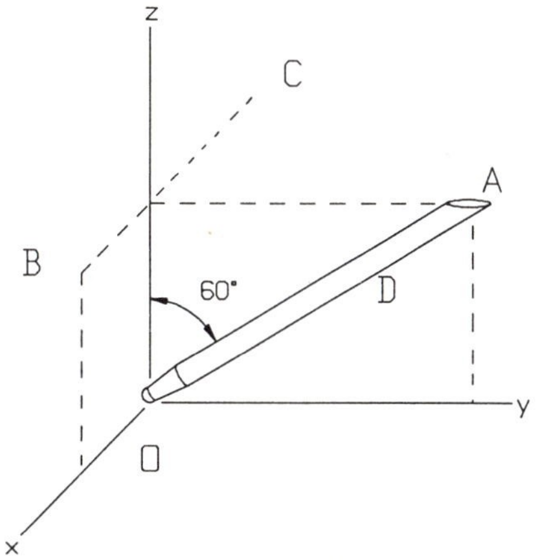

2. <u>Identify the unknowns and count the number of independent equilibrium equations.</u>

✍ (a) List the unknowns on the FBD:

✍ (b) Number of independent equilibrium equations =

Method of Analysis

Since three of the unknowns, the components of the ball-and-socket reaction, pass through point O, the equation $\Sigma \mathbf{M}_O = \mathbf{0}$ will contain the two unknown tensions. Therefore, the components (**i** and **j**) of $\Sigma \mathbf{M}_O = \mathbf{0}$ can be used to solve for T_{AC} and T_{AB}. The components of the equation $\Sigma \mathbf{F} = \mathbf{0}$ can be used to solve for O_x, O_y, and O_z.

3. Write and solve the equations to determine T_{AC} and T_{AB}.

✍ (a) Write the forces in vector form:

$\mathbf{T}_{AC} =$

$\mathbf{T}_{AB} =$

Letting **P** be the 5000-lb force: **P** =

✍ (b) Write the position vector from O to A: $\mathbf{r}_{OA} =$

Write the position vector from O to D: $\mathbf{r}_{OD} =$

✍ (c) Write and solve the equation $\Sigma \mathbf{M}_O = \mathbf{0}$:

$(\mathbf{r}_{OA} \times \mathbf{T}_{AC}) + (\mathbf{r}_{OA} \times \mathbf{T}_{AB}) + (\mathbf{r}_{OD} \times \mathbf{P}) = \mathbf{0}$

$$\begin{vmatrix} \mathbf{i} & \mathbf{j} & \mathbf{k} \\ & & \\ & & \end{vmatrix} + \begin{vmatrix} \mathbf{i} & \mathbf{j} & \mathbf{k} \\ & & \\ & & \end{vmatrix} + \begin{vmatrix} \mathbf{i} & \mathbf{j} & \mathbf{k} \\ & & \\ & & \end{vmatrix} = \mathbf{0}$$

Equate the like components:

Σ(**i** components) = 0:

Σ(**j** components) = 0:

Solve for T_{AC} and T_{AB}:

4. Determine the resultant reaction at O.

✍ (a) Solve for O_x: $\Sigma F_x = 0$:

✍ (b) Solve for O_y: $\Sigma F_y = 0$:

✍ (c) Solve for O_Z: $\Sigma F_Z = 0$:

✍ (d) The resultant reaction at O (in vector form): $\mathbf{R}_O =$

$T_{AC} = 3070$ lb
$T_{AB} = 3950$ lb
$\mathbf{R}_O = 1.47\mathbf{i} + 5780\mathbf{j} + 5000\mathbf{k}$ lb

Another Method of Analysis

The equation $\Sigma M_{OB} = 0$ can be used to solve for T_{AC}, and $\Sigma M_{OC} = 0$ can be used to solve for T_{AB}.

S 24. FRICTION PROBLEM CLASSIFICATION AND ANALYSIS

Text Reference: Articles 7.2, 7.3; Sample Problems 7.1 - 7.7

A. *You Should Understand:*

- A friction force is the component of a resultant contact reaction force that is tangent to a rough surface.
- F_{max} is the maximum static friction force that can exist between two surfaces in contact that are not moving relative to each other.
- The coefficient of static friction, μ_s, is an experimental constant ($F_{max} = \mu_s N$, where N is the component of a resultant contact reaction force that is perpendicular to a surface).
- The friction force F is always less than or equal to F_{max} ($F \leq F_{max}$). In other words, $F = \mu_s N$ is not always true. Therefore, a single method of analysis is not sufficient to handle all possible friction problems.
- If $F = F_{max}$, the surfaces are on the verge of sliding relative to each other. This condition is called impending sliding.
- The direction of F_{max} is opposite that in which the surface would tend to slide.
- The friction force between two surfaces that are sliding relative to each other is called kinetic or dynamic friction, idicated as F_k.
- The coefficient of kinetic friction, μ_k, is an experimental constant, usually smaller than μ_s ($F_k = \mu_k N$).
- The direction of F_k opposes the relative sliding.
- Friction problems involving sliding can be classified into three types, as follows:

  Type I: A problem in which impending sliding is not specified.

  Type II: A problem in which impending sliding is implied and the surface(s) where sliding impends is (are) known.

  Type III: A problem in which impending sliding is implied but the surface(s) where sliding impends is (are) not known.

B. *You Should be Able to:*

- Draw free-body diagrams of bodies where rough surfaces are involved.
- Classify friction problems according to their type and solve them (Article 7.3 of the text describes the method of analysis for each type of friction problem).

C. Guided Practice Problems

S 24.1 The 30-kg block A is at rest in the position shown when the force $P = 100$ N is applied to it. The coefficient of friction between block A and the horizontal plane is 0.4. Detemine the value of the friction force acting on block A.

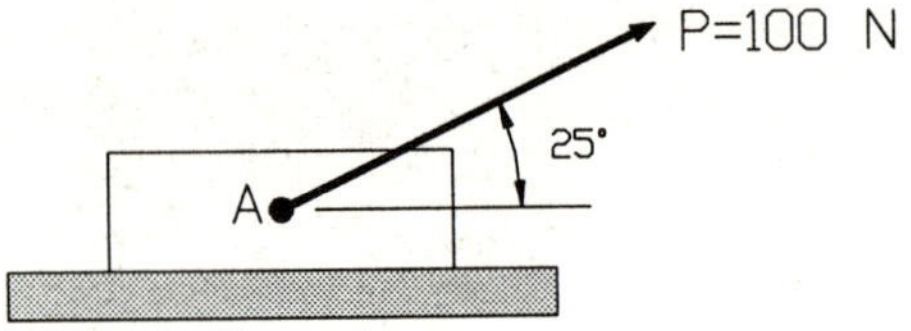

GUIDED SOLUTION

The first step in this friction problem is to identify the type of problem. Since we do not know whether sliding impends, this is a Type I problem.

1. Assume equilibrium. The assumption implies that P is not sufficient to move the block.

2. Draw the FBD of block A. Label the normal force at the surface N_A and the friction force F_A.

3. Write and solve the equilibrium equations for the friction and normal forces. (Remember that equilibrium has been assumed.)

   (a) Solve for N_A: $\Sigma F_y = 0$:

   (b) Solve for F_A: $\Sigma F_x = 0$:

4. Check if $F_A \le F_{max}$. If $F_A \le F_{max}$, then the assumption of equilibrium was correct and the friction force under the block is F_A.

   $F_{max} = \mu_s N_A = 0.4 N_A =$

   Is $F_A \le F_{max}$?

$F_A = 90.6$ N

S 24.2 What is the minimum weight of block A for which the system will remain at rest? The weight of block B is 600 lb. For the inclined surface under block A, $\mu_s = 0.2$ and for the horizontal surface under block B, $\mu_s = 0.12$.

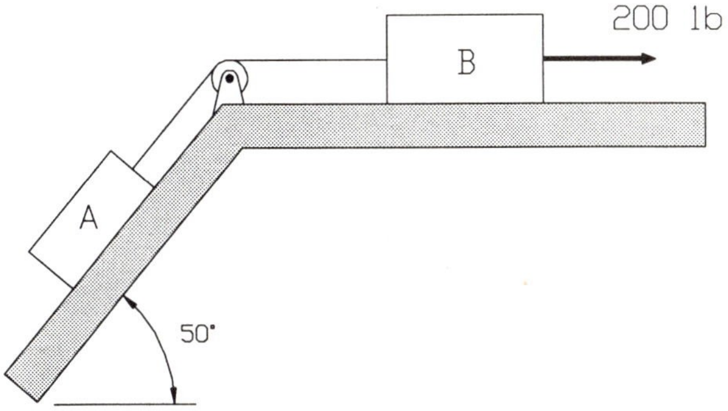

GUIDED SOLUTION

What type of problem is this? Do we know if sliding impends? The statement, 'the minimum weight of block A for which the system will remain at rest' implies impending sliding. Therefore, this problem is not a Type I problem. Do we know the suface or surfaces where sliding impends? Because the two blocks are connected by a cable, one block cannot slide without the other sliding. Consequently, we know that sliding impends for both blocks. Therefore, this is a Type II problem.

Unnecessary complications can be avoided if the correct directions are used for the friction forces. The direction of the friction force on each block is opposite the direction of impending sliding. In this case, there are two possibilities for the direction of impending sliding. If block A is too heavy, it will slide down the incline. If it is too light, block A will slide up the incline. Therefore, the lightest weight of block A corresponds to impending motion of A upward and impending motion of B to the right.

1. Draw the FBD of each block and set $F = F_{max}$ for both blocks. (Be sure to show friction in the correct direction.)

✍

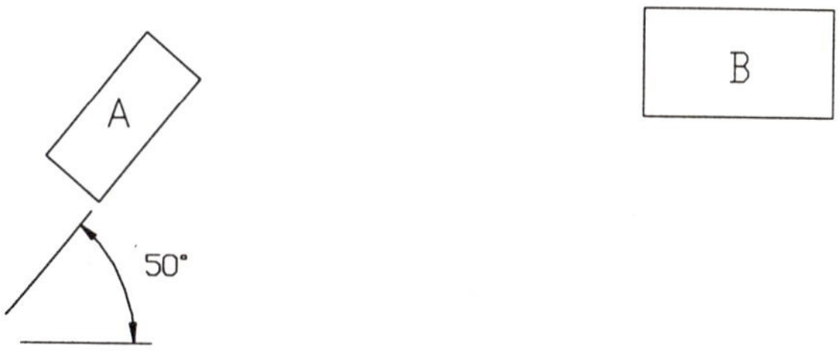

2. Compare the number of unknowns on the two free-body diagrams with the number of equilibrium equations.

✍ List the unknowns of the FBDs:

The number of available equilibrium equations =

3. Write the equilibrium equations and solve for the weight of A.

✍

$W_A = 143$ lb

S 24.3 Block A has a mass of 60 kg and block B has a mass of 80 kg. The coefficient of static friction between the blocks (surface ①) is 0.45 and between block B and the inclined plane (surface ②), $\mu_s = 0.6$. Determine the angle θ for which motion impends.

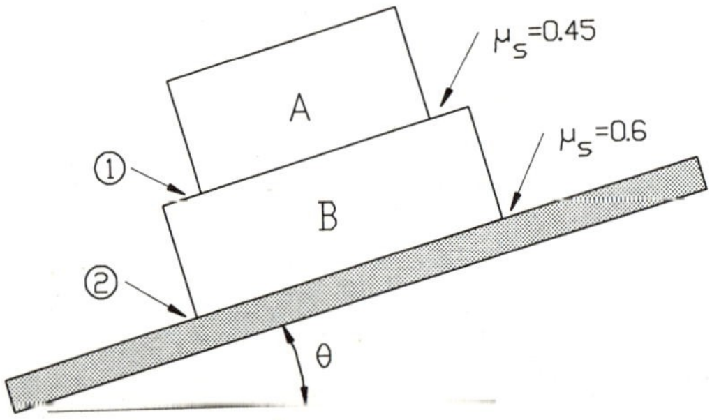

GUIDED SOLUTION

What type of problem is this? Is impending sliding specified or implied? Since the problem asks for the angle for which motion impends, impending sliding is obviously specified. Therefore, this is not a Type I problem. Do we know the suface or surfaces where sliding impends? There are two possible surfaces where sliding can impend: surface ① or surface ②. Since we do not know on which surface sliding impends, this is a Type III problem.

We will solve this problem using the two methods described in the text for a Type III problem.

Method I

1. Assume that sliding impends at surface ①. The assumption means that F_1, the friction force at surface ① equals $(F_1)_{max}$. The direction in which motion impends (the direction in which A will tend to slide if B were fixed) must be determined so that F_1 can be drawn in the correct direction. The direction of the other friction force, F_2, can be assumed.

2. <u>Draw the FBD of both blocks</u>. Assume F_2 is acting to the right and call the normal forces at surfaces ① and ②, N_1 and N_2, respectively.

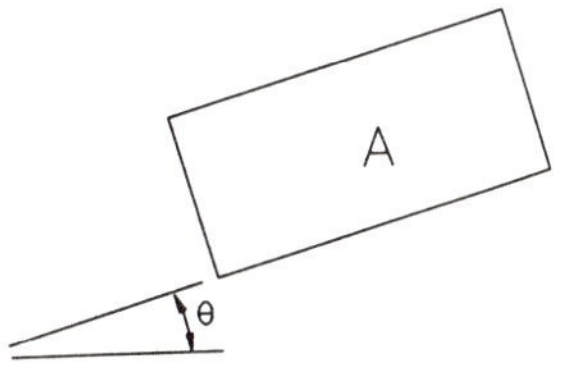

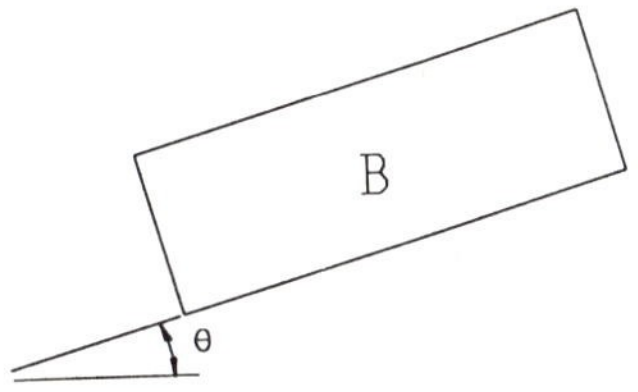

Note that there are really 5 unknowns on the two FBDs: N_1, F_1, N_2, F_2, and θ. We have 4 equilibrium equations and one equation from the assumption that $F_1 = (F_1)_{max} = \mu_s N_1$.

3. <u>Write the equilibrium equations and solve for θ</u>.

4. <u>Assume that sliding impends at surface ②</u>. The assumption means that F_2, the friction force at surface ② is set as $(F_2)_{max}$. The direction in which motion impends (the direction in which B will tend to slide on the inclined plane) must be determined so that F_2 can be drawn in the correct direction on B. The direction of the other friction force, F_1, can be assumed.

5. <u>Draw the FBD of both blocks.</u>

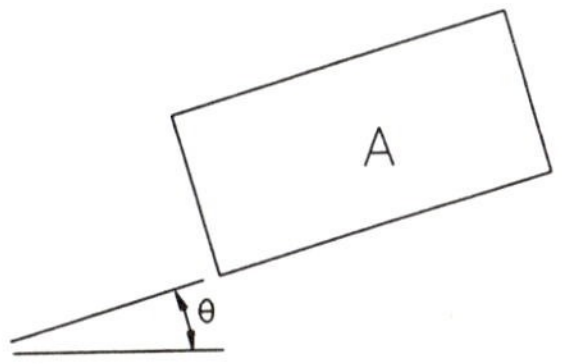

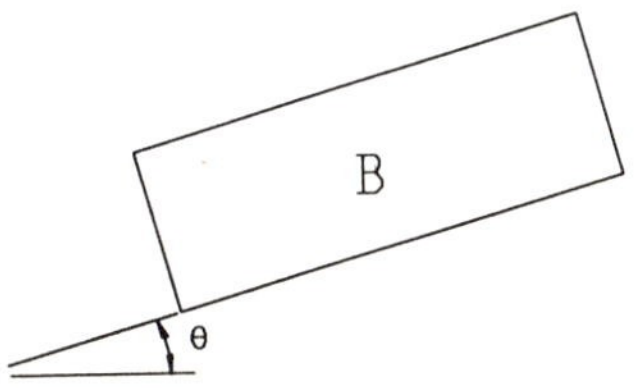

6. Write the equilibrium equations and solve for θ.

✍

7. Comparing the results of Steps 3 and 6, choose the correct angle.

✍ $\theta =$

$\theta = 24.2°$

Method II

The first three steps for both methods are the same and will not be repeated here in Method II.

1. Using the FBDs in step 2 above, write the equilibrium equations and solve for all the unknowns.

✍

2. Check if $F_2 \leq F_{max}$. If $F_2 \leq F_{max}$, the angle determined above (Method I) is the correct answer. Otherwise, steps 4, 5, and 6 of Method I must be repeated.

✍ $F_{max} = \mu_s N = 0.6N_2 =$

Is $F_2 \leq F_{max}$?

$\theta =$

S 25. IMPENDING TIPPING

Text Reference: Article 7.4; Sample Problems 7.8 - 7.10

A. *You Should Understand:*

- If tipping impends, the line of action of the normal reaction force is at a corner of the body about which the body is about to tip.
- If tipping does not impend, the location of the line of action of the normal reaction force is unknown.
- The three classifications of friction problems discussed in the previous section can be modified to include the possibility of tipping. Whenever 'sliding' occurs in the description of the problem types, the term 'motion' (sliding or tipping) is substituted.

B. *You Should be Able to:*

- Analyze problems to determine whether impending sliding and/or tipping is (are) possible.
- Classify problems that involve impending tipping as well as impending sliding according to their type and solve them.
- Draw the normal force in the correct position if impending tipping occurs or is assumed to occur.

C. Guided Practice Problems

S 25.1 Determine whether sliding or tipping impends as the angle θ is gradually increased from zero. The mass of the block is 10 kg and at the surface between the block and the plane, $\mu_s = 0.70$.

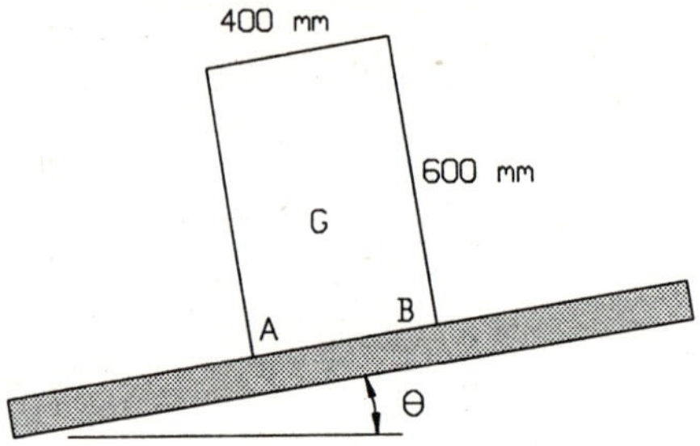

GUIDED SOLUTION

What type of problem is this? Do we know if motion (sliding or tipping) impends? In this case, impending motion is clear from the statement of the problem. Therefore, this is not a Type I problem. However, we cannot determine from observation whether sliding or tipping impends. Consequently, this is a Type III problem.

We will solve this problem using Method I for a Type III problem.

1. Assume impending sliding and draw the FBD of the block. The assumption means that F, the friction force, is set as F_{max}. Be sure to draw F in the correct direction. (G is the center of gravity of the block.

✍

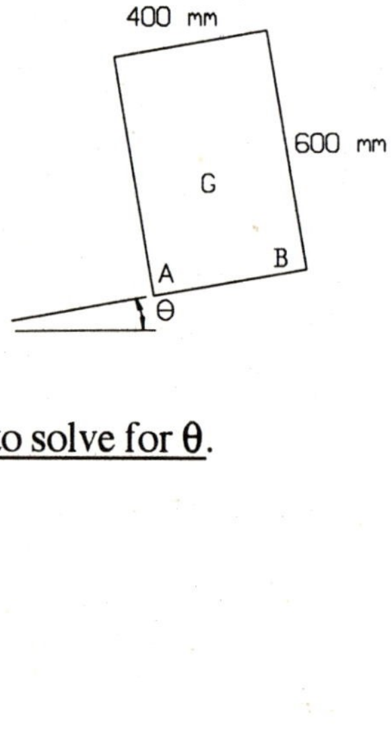

2. Write the equilibrium equations to solve for θ.

✍

3. <u>Assume impending tipping and draw the FBD of the block.</u> The assumption means that N, the normal force acts at the corner about which tipping of the block impends.

✍

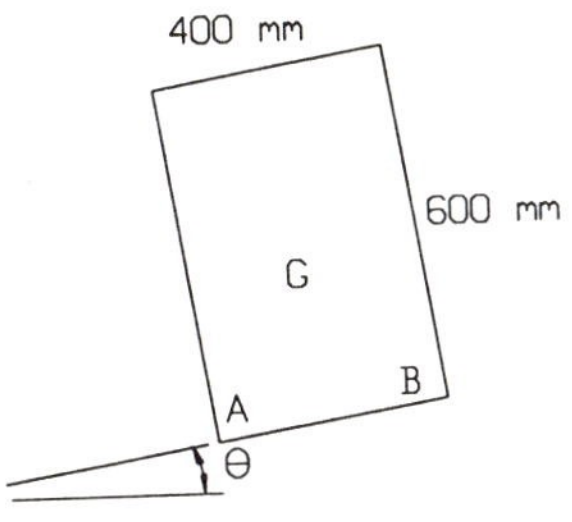

4. <u>Write the equilibrium equations to solve for θ.</u>

✍

6. <u>Choose correct type of impending motion.</u>

✍ Type of motion that impends: at θ =

Tipping at θ = 33.7°

Note: For Method II , the location of the line of action of the normal force would be determined as step 3 of this problem. If the line of the normal force was determined to act beyond the corner of the block, impending tipping before sliding would be indicated. The angle for impending tipping would then have to be determined from the FBD of the impending tipping condition.

S 26. CENTROIDS OF PLANE AREAS AND CURVES

Text Reference: Article 8.2; Sample Problems 8.1 - 8.4

A. *You Should Understand:*

- The centroid is the geometric center of an area or a curve.
- Q_x and Q_y are the first moments of an area or a curve about the x and y axes, respectively. The first moments are used to define the coordinates of the centroid of an area or a curve.
- If the coordinates $(\bar{x},\bar{y})$ of the centroid of an area or a curve are known, $Q_x = A\,\bar{y}$ and $Q_y = A\,\bar{x}$ for the area A, or $Q_x = L\bar{y}$ and $Q_y = L\,\bar{x}$ for the curve of length L.
- The methods of composite areas and composite curves are used to locate the centroid of a composite shape or a composite curve, respectively.

B. *You Should be Able to:*

- Use single or double integration to find the area of a plane region and the first moments of an area about the x and y axes.
- Calculate the coordinates of the centroid of a plane region using the first moments of the area and the area of the plane region.
- Using integration, determine the length of a plane curve and the first moments of a curve about the x and y axes.
- Calculate the coordinates of the centroid of a plane curve using the first moments of the curve and the length of the plane curve.
- Use the methods of composite areas and composite curves to determine the coordinates of the centroid of a composite shape and the coordinates of the centroid of a composite curve.

C. Guided Practice Problems

S 26.1 Determine the coordinates $(\bar{x},\bar{y})$ of the centroid of the shaded area using the following methods:

(a) Double integration.
(b) Single integration with a vertical element.
(c) Single integration with a horizontal element.

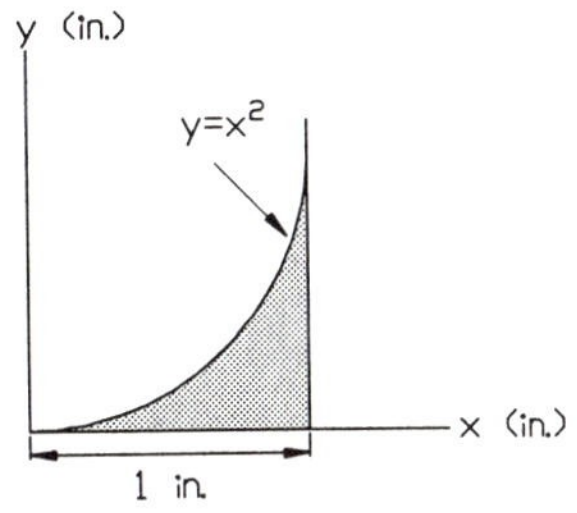

GUIDED SOLUTION

Part (a)

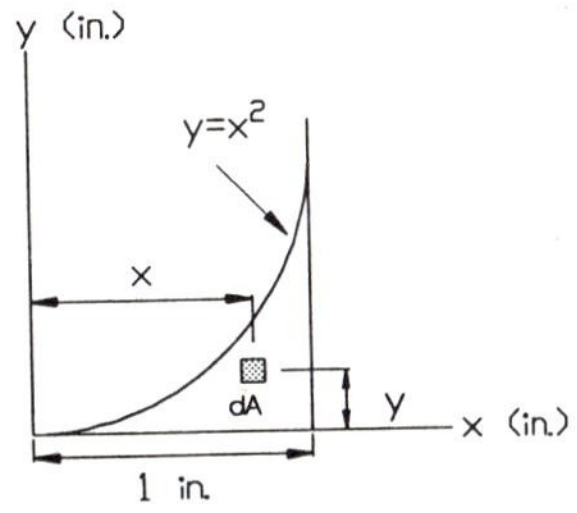

1. Using the differential element of area dA, shown above, determine the limits of integration for the calculation of the total area and determine the area.

✍ $dA = dy\,dx$

$$A = \int \int dy\,dx =$$

2. Determine the first moment about the x axis.

✍ $dQ_x = y\,dy\,dx$

$$Q_x = \int_0^1 \int_0^{x^2} y\,dy\,dx =$$

3. Determine the first moment about the y axis.

✍ $dQ_y = x\,dy\,dx$

$$Q_y = \int_0^1 \int_0^{x^2} x\,dy\,dx =$$

4. Calculate the coordinates of the centroid.

$\bar{y} = Q_x / A =$

$\bar{x} = Q_y / A =$

$\bar{x}$= 0.750 in., $\bar{y}$= 0.300 in.

Part (b)

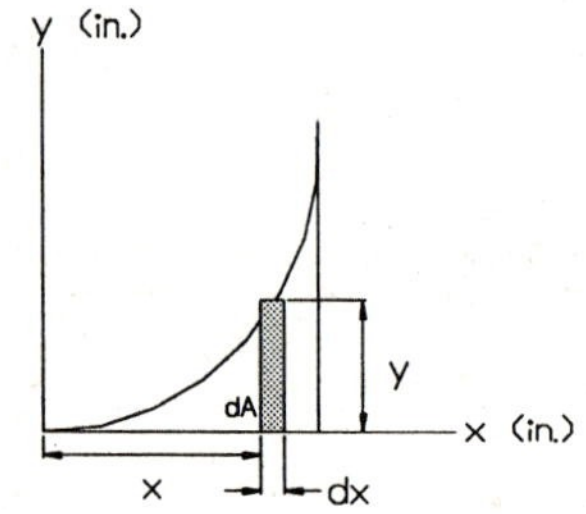

1. Determine the total area.

$dA = y\,dx = x^2 dx$

$A = \int_0^1 x^2 dx =$

2. Determine the first moment about the x axis. The distance from the center of the vertical element to the x-axis is y/2.

$dQ_x = y/2\ dA$

$Q_x = \int_0^1 \frac{y}{2}\, dA = \int_0^1 \frac{x^4}{2}\, dx =$

3. Determine the first moment about the y axis.

$dQ_y = x\, dA$

$Q_y = \int_0^1 x\, dA = \int_0^1 x^3\, dx =$

4. Calculate the coordinates of the centroid.

$\bar{y} = Q_x / A =$

$\bar{x} = Q_y / A =$

Part (c)

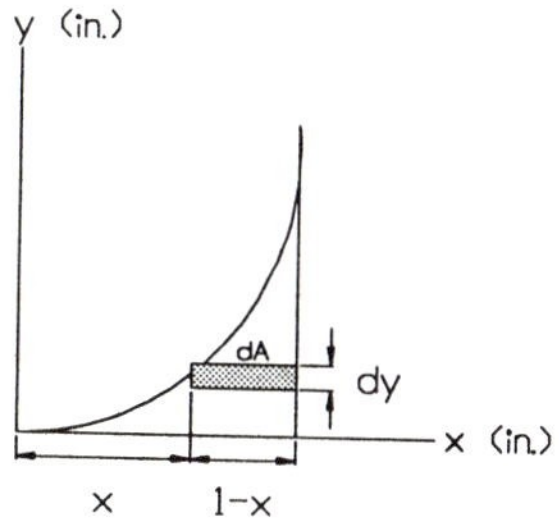

1. Determine the total area. Note that at x = 1, y = 1.

✍ $dA = (1-x)\,dy = (1-\sqrt{y}\,)\,dy$

$$A = \int_0^1 (1-\sqrt{y})\,dy =$$

2. Determine the first moment about the x axis.

✍ $dQ_x = y\,dA = y\,(1-\sqrt{y}\,)\,dy$

$$Q_x = \int_0^1 y(1-\sqrt{y})\,dy =$$

3. Determine the first moment about the y axis. The distance from the center of the horizontal strip to the y axis is [x + (1–x)/2].

✍ $$dQ_y = x + \frac{1-x}{2}dA = \frac{1+x}{2}dA = \frac{1+\sqrt{y}}{2}(1-\sqrt{y})dy$$

$$Q_y = \int_0^1 \left(\frac{1+\sqrt{y}}{2}\right)(1-\sqrt{y})\,dy =$$

4. Calculate the coordinates of the centroid.

✍ $\bar{y} = Q_x / A =$

$\bar{x} = Q_y / A =$

S 26.2 Using the method of composite areas, determine the coordinates $(\bar{x},\bar{y})$ of the centroid of the shaded area.

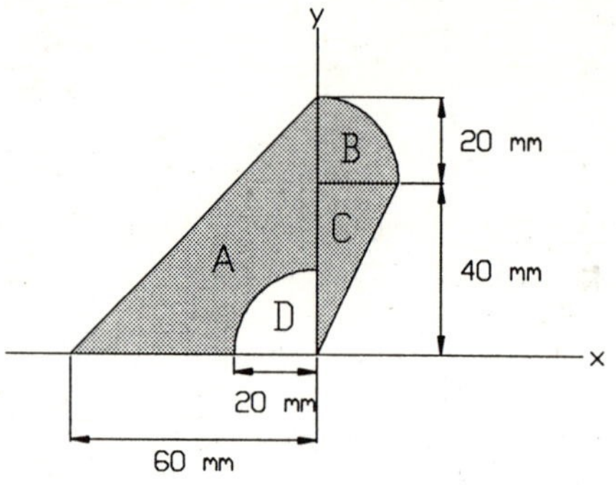

GUIDED SOLUTION

1. Complete the following table to calculate the total area and the first moments of the parts about the x and y axes. Use Table 8.1 in the text to locate the centroids of the simple shapes.

✍

Part	Area (mm^2)	$\bar{x}$ (mm)	$Q_y = A\bar{x}$ (mm^3)	$\bar{y}$ (mm)	$Q_x = A\bar{y}$ (mm^3)
A. Large triangle					
B. Quarter circle					
C. Small triangle					
D. Quarter circle space					
	$\Sigma A =$		$\Sigma(A\bar{x})=$		$\Sigma(A\bar{y})=$

2. Calculate the coordinates of the centroid.

✍ $\bar{x} = \Sigma Q_y/\Sigma A = \Sigma(A\bar{x})/\Sigma A =$

$\bar{y} = \Sigma Q_x/\Sigma A = \Sigma(A\bar{y})/\Sigma A =$

$\bar{x} = -12.7$ mm, $\bar{y} = 26.9$ mm

S 27. Centroids of Curved Surfaces, Volumes and Space Curves

Text Reference: Article 8.3; Sample Problems 8.5 - 8.10

A. *You Should Understand:*

- Three coordinates are required to locate the centroid of three-dimensional shapes.
- If a region has a plane of symmetry, its centroid lies in that plane.
- The method of composite shapes can be applied to curved surfaces, volumes, and space curves.

B. *You Should be Able to:*

- Locate the centroid of surfaces, volumes and space curves.
- Use the method of composite shapes to determine the coordinates of its centroid using Tables 8.3 and 8.4 in the text.

C. Guided Practice Problems

S 27.1 A 4-in. diameter hole is drilled into a 12 x 12 x 9-in. solid block, as shown. The hole is 6 in. deep. Determine the coordinates of the center of the volume using the method of composite shapes.

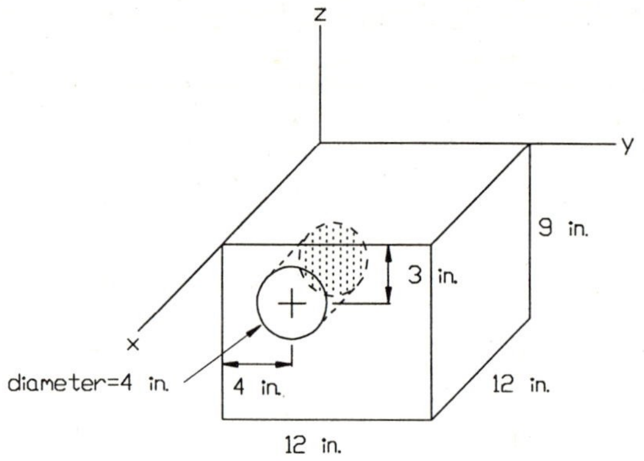

GUIDED SOLUTION

1. Complete the following table to calculate the total volume and the first moments of the parts about the x, y and z axes.

Part	Vol. (in.^{3})	$\bar{x}$ (in.)	$V\bar{x}$ (in.^{4})	$\bar{y}$ (in.)	$V\bar{y}$ (in.^{4})	$\bar{z}$ (in.)	$V\bar{z}$ (in.^{4})
Block							
Cylindrical hole							
	$\Sigma V =$		$\Sigma V\bar{x} =$		$\Sigma V\bar{y} =$		$\Sigma V\bar{z} =$

2. Calculate the coordinates of the centroid of the block.

$\bar{x} = \Sigma V\bar{x} / \Sigma V =$

$\bar{y} = \Sigma V\bar{y} / \Sigma V =$

$\bar{z} = \Sigma V\bar{z} / \Sigma V =$

$\bar{x}$ = 5.81 in., $\bar{y}$ = 6.12 in., $\bar{z}$ = – 4.59 in.

S 28. THEOREMS OF PAPPUS-GULDINUS

Text Reference: Article 8.4; Sample Problem 8.11

A. *You Should Understand:*

- The Theorems of Pappus-Guldinus are convenient for calculating the surface areas and volumes of bodies of revolution.

B. *You Should be Able to:*

- Determine the surface area of a shape generated by a revolving body using Theorem I.
- Determine the volume of a shape generated by a revolving body using Theorem II.

C. Guided Practice Problems

S 28.1 Compute the surface area generated by revolving the lines shown about the y axis.

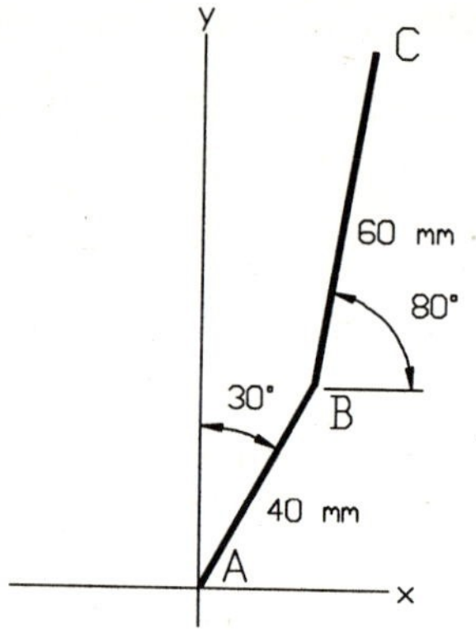

GUIDED SOLUTION

The total surface area is the sum of the surface areas generated by the rotation of each of the lines.

1. Determine the x coordinate of the centroid of each line.

✍ For segment AB: $\bar{x} = 1/2\,\overline{AB}\sin 30° =$

For segment BC: $\bar{x} = \overline{AB}\sin 30° + 1/2\,\overline{BC}\sin 10° =$

2. Determine the first moment of each line about the y axis.

✍ For segment AB: $Q_x = L\bar{x} =$

For segment BC: $Q_x = L\bar{x} =$

3. Determine the surface area generated by each line and sum. (Theorem I)

✍ For segment AB: $A = 2\pi Q_x =$

For segment BC: $A = 2\pi Q_x =$

Total Area =

Area = 12.02×10^3 mm^2

S 29. Center of Gravity and Center of Mass

Text Reference: Article 8.5; Sample Problem 8.12

A. *You Should Understand:*

- The weight of a body is a distributed force. The total weight of a body is the resultant of the distributed force and acts through the center of gravity of the body.
- The center of gravity and the center of mass coincide, unless the gravitational field is not constant.
- The weight density γ is constant for a homogeneous body. Therefore, the center of gravity coincides with the centroid of the volume of the body.
- The weight density γ and the mass density ρ are related by $\gamma = \rho g$, where g is the local acceleration due to gravity.

B. *You Should be Able to:*

- Determine the location of the center of gravity of a body.
- Locate the center of gravity of a composite body.

C. *Guided Practice Problems*

S 29.1 A 12 x 12 x 9-in. wood block weighs 40 lb/ft^3. The block contains a 4-in. diameter hole that is 6 in. deep and is filled with a solid metal cylinder that weighs 1146 lb/ft^3. Determine the coordinates of the center of gravity of the composite body.

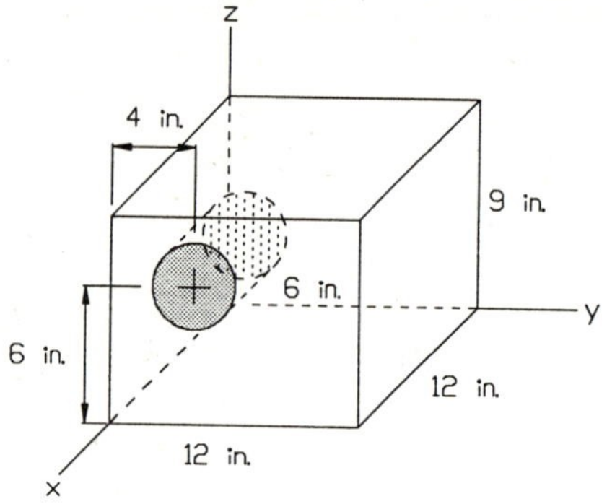

GUIDED SOLUTION

We consider the body to be composed of three parts: a solid wood block, the wood cylindrical hole (to be subtracted from the block), and the metal cylinder. Since each part is homogeneous, its center of gravity coincides with the centroid of its volume.

1. Determine the weight of the parts in pounds.

✍ Wood block: Weight = γV =

Wood cylinder: Weight = γV =

Metal cylinder: Weight = γV =

2. Complete the following table to calculate the total weight and the first moments of the weights about the x, y, and z axes.

✍

Part	Weight (lb)	$\bar{x}$ (in.)	$W\bar{x}$ (lb•in.)	$\bar{y}$ (in.)	$W\bar{y}$ (lb•in.)	$\bar{z}$ (in.)	$W\bar{z}$ (lb•in.)
Wood Block							
Wood cylinder							
Metal cylinder							
	ΣW =		$\Sigma W\bar{x}$=		$\Sigma W\bar{y}$=		$\Sigma W\bar{z}$=

3. Calculate the coordinates of the center of gravity of the body.

✍ $\bar{x} = \Sigma W\bar{x} / \Sigma W$ =

$\bar{y} = \Sigma W\bar{y} / \Sigma W$ =

$\bar{z} = \Sigma W\bar{z} / \Sigma W$ =

$\bar{x}$= 7.85 in., $\bar{y}$= 4.77 in., $\bar{z}$= 5.42 in.

S 30. Distributed Normal Loads

Text Reference: Article 8.6; Sample Problems 8.13 - 8.15

A. *You Should Understand:*

- The resultant of a normal load on a flat surface equals the volume of the region between the load surface and the load area, and passes through the centroid of the volume.
- The resultant of a line load equals the area of the region under the load diagram and passes through the centroid of the region.
- Uniform pressure is the special case in which the magnitude of the load intensity is constant.
- Equilibrium analysis can be used to determine the resultant force due to fluid pressure.

B. *You Should be Able to:*

- Determine the resultant of a normal load on a flat surface and its line of action.
- Determine the resultant of a line load and its line of action.
- Determine the resultant of a uniform pressure load and its line of action.
- Determine the resultant of a fluid pressure load and its line of action.

C. Guided Practice Problems

S 30.1 A tank of rectangular cross section is filled with water 1.2 m deep, as shown. The density of the water is 1000 kg/m^3. Determine the magnitude of the resultant force exerted by the water on the right vertical wall and the coordinates of the point where the resultant intersects the right vertical wall.

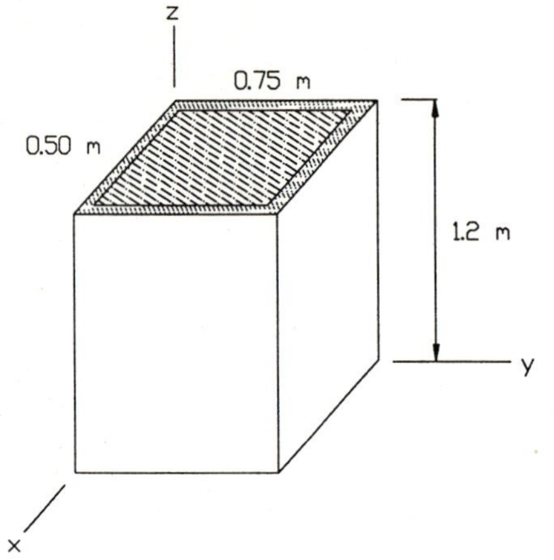

GUIDED SOLUTION

By symmetry, the x coordinate of the point where the resultant intersects the right wall is 0.25 m.

1. Determine the weight density of water.

✍ $\gamma = \rho g =$

2. Determine the water pressure at the bottom of the tank. The pressure varies linearly from zero at the top to the maximum at the bottom.

✍ p_{bottom} = water depth x γ =

3. The pressure distribution on the right vertical wall is shown.

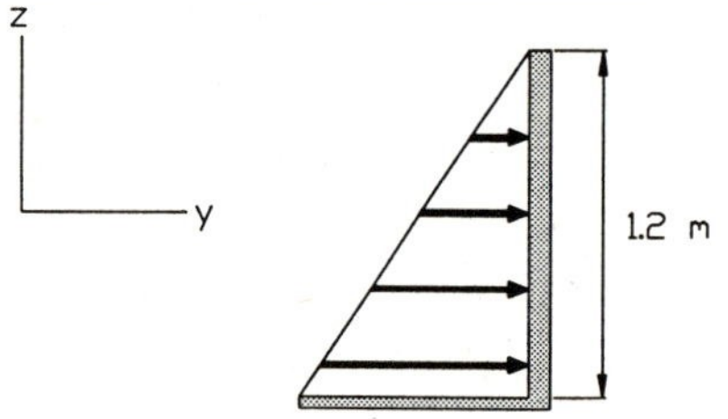

4. Determine the resultant force applied by the pressure region. Since the pressure distribution is triangular, the average pressure is used to determine the resultant force that is applied over the entire right vertical wall which is 0.50 m wide.

✍ Resultant Force =

5. Determine the z coordinate of the line of action of the resultant force. The resultant acts at the centroid of the triangular area.

✍ z =

Force = 3530 N, z = 0.40 m

S 31. Moment of Inertia of Areas; Polar Moment of Inertia

Text Reference: Article 9.2; Sample Problems 9.1 - 9.5

A. *You Should Understand:*

- The moment of inertia of the area about an axis is also called the second moment of the area about an axis.
- The moment of inertia and the polar moment of an area are always positive.
- The parallel-axis theorem uses the moment of inertia about a centroidal axis to calculate the moment of inertia about a parallel axis.
- The radius of gyration is determined by computation, $\sqrt{I / A}$ or $\sqrt{J / A}$.
- The method of composite areas can be used to determine the moment of inertia of areas composed of two or more common shapes.
- Tables 9.1 and 9.2 in the text list the inertial properties for common geometric shapes.

B. *You Should be Able to:*

- Determine the moment of inertia of an area by integration.
- Determine the polar moment of an area by integration.
- Use the parallel-axis theorem.
- Calculate the radius of gyration about an axis.
- Use the method of composite areas.

C. Guided Practice Problems

S 31.1 For the shaded area shown, determine the following:

(a) The moment of inertia about the x axis using double integration.
(b) The moment of inertia about the x axis using single integration with a horizontal element.
(c) The moment of inertia about the x axis using single integration with a vertical element.
(d) The radius of gyration with respect to the x axis.

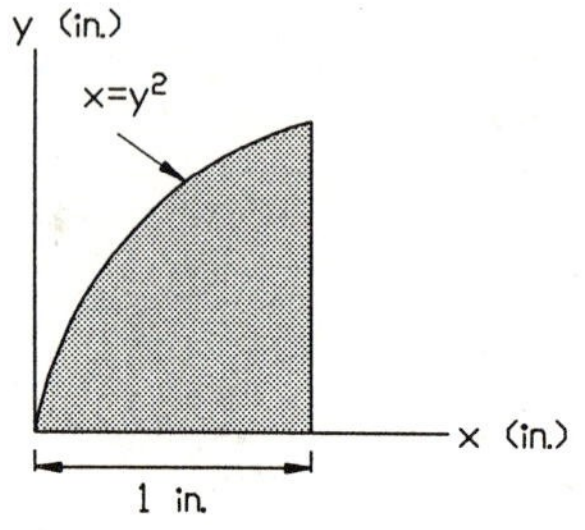

GUIDED SOLUTION

Part (a)

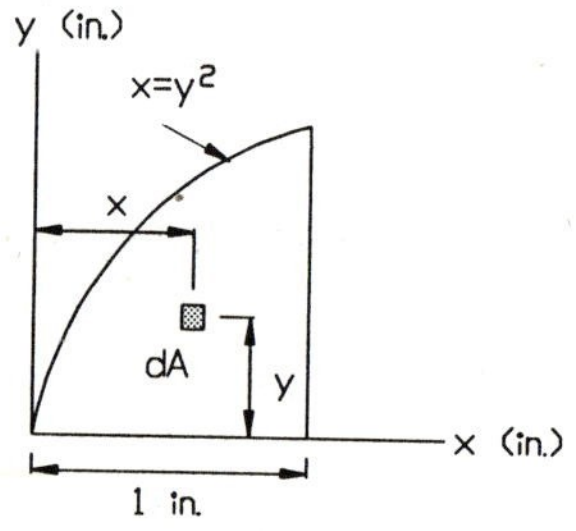

1. Calculate I_x.

$$dA = dy\,dx$$

✍ $I_x = \int_0^1 \int_0^{\sqrt{x}} y^2 dy dx =$

$\boxed{I_x = 0.133 \text{ in.}^4}$

Part (b)

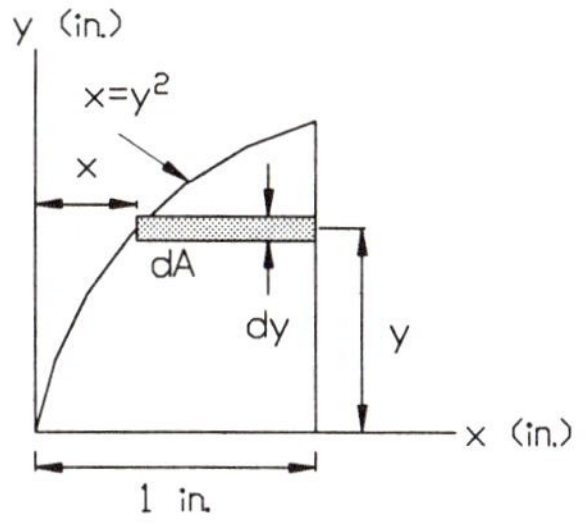

1. <u>Calculate</u> I_x.

$$dA = (1-x)\,dy = (1-y^2)\,dy$$

✍ $$I_x = \int_0^1 y^2\left(1-y^2\right)dy =$$

Part (c)

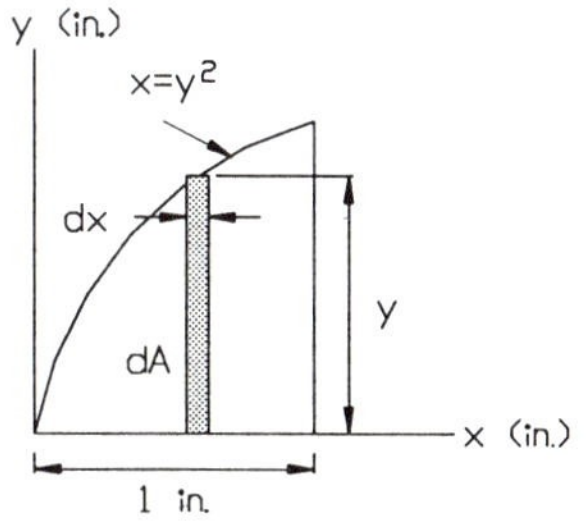

1. <u>Calculate</u> I_x. From Table 9.1 in the text, we have $I_x = bh^3/3$ for a rectangular area.

$$dI_x = dx(y^3/3) = dx\frac{x^{3/2}}{3}$$

✍ $$I_x = \int_0^1 \frac{x^{3/2}}{3}dx =$$

Part (d)

1. <u>Calculate</u> k_x.

✍ (a) Calculate the total area: $A = \int_0^1 \int_0^{\sqrt{x}} dy\,dx =$

✍ (b) Calculate the radius of gyration: $k_x = \sqrt{\frac{I_x}{A}} =$

$k_x = 0.447$ in.

S 31.2 Using the method of composite areas, calculate the moment of inertia of the shaded area about the x and y axes.

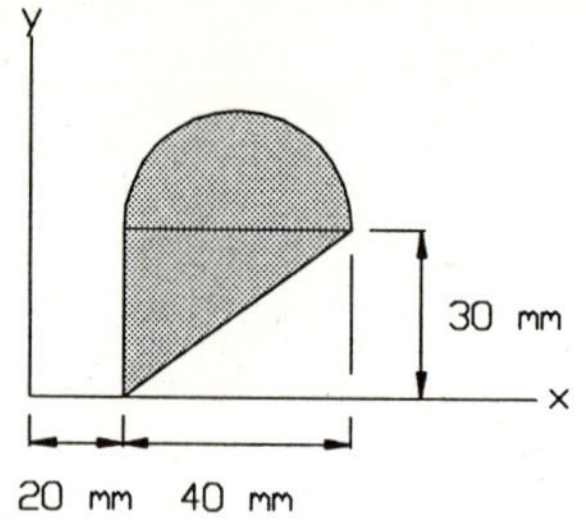

GUIDED SOLUTION

From Table 9.1 of the text, we can obtain the inertial properties of a semicircle and a triangle. Since neither the x- nor the y-axis is centroidal to the two shapes composing the area, we will need to use the parallel-axis theorem to determine I_x and I_y.

1. For the triangle, determine the area, I_x and I_y.

(a) $A = 1/2(bh) =$

(b) $I_x = \bar{I}_x + Ad^2 = \dfrac{bh^3}{36} + Ad^2 =$

(c) $I_y = \bar{I}_y + Ad^2 = \dfrac{hb^3}{36} + Ad^2 =$

2. For the semicircle, determine the area, I_x and I_y.

(a) $A = \pi R^2/2 =$

(b) $I_x = \bar{I}_x + Ad^2 = 0.1098R^4 + Ad^2 =$

(c) $I_y = \bar{I}_y + Ad^2 = \dfrac{\pi R^4}{8} + Ad^2 =$

3. Sum the moments of inertia for the two shapes.

$I_x = \Sigma I_x =$

$I_y = \Sigma I_y =$

$I_x = 1.218 \times 10^6 \text{ mm}^4$, $I_y = 1.788 \times 10^6 \text{ mm}^4$

S 32. PRODUCTS OF INERTIA OF AREAS

Text Reference: Article 9.3; Sample Problems 9.6 - 9.8

A. *You Should Understand:*

- The product of inertia of an area can be positive, negative or zero.
- The product of inertia is zero relative to a set of axes if one of the axes is an axis of symmetry. The set of axes consists of the axis of symmetry as well as the axis perpendicular to it.
- Products of inertia obey a parallel-axis theorem.
- The product of inertia of an area composed of two or more common shapes can be determined by the method of composite areas.
- Tables 9.1 and 9.2 in the text list the inertial properties for common geometric shapes.

B. *You Should be Able to:*

- Determine the product of inertia of an area about coordinate axes.
- Use the parallel-axis theorem for products of inertia of areas.
- Use the method of composite areas for the product of inertia.

C. Guided Practice Problems

S 32.1 Using the method of composite areas, calculate the product of inertia I_{xy} for the shaded area shown.

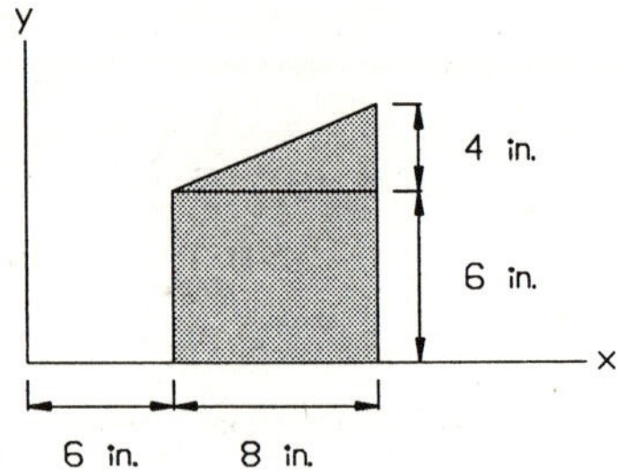

GUIDED SOLUTION

The parallel-axis theorem must be used for the rectangle and triangle that make up the shaded area. The shaded area could also be treated as a rectangle (10 in. x 8 in.) minus a right triangle. Note that by symmetry, $\bar{I}_{xy} = 0$ for the rectangle.

1. Calculate I_{xy} for the rectangle.

✍ $I_{xy} = \bar{I}_{xy} + A\bar{x}\bar{y} =$

2. Calculate I_{xy} for the triangle. Use Table 9.1 for the inertial properties of a right triangle.

✍ $I_{xy} = \bar{I}_{xy} + A\bar{x}\bar{y} =$

3. Calculate I_{xy} for the area composed of the rectangle and triangle.

✍ $I_{xy} = \Sigma I_{xy} =$

$I_{xy} = 2760 \text{ in.}^4$

S 33. TRANSFORMATION EQUATIONS; PRINCIPAL MOMENTS OF INERTIA OF AREAS

Text Reference: Article 9.4; Sample Problem 9.9

A. *You Should Understand:*

- The transformation equations describe the inertial properties as functions of the angle through which a coordinate system is rotated.
- The principal moments of inertia at a point are the maximum and minimum moments of inertia at the point.
- The principal axes are the axes about which the moments of inertia are maximum or minimum. The directions of the principal axes are called the principal directions.
- The product of inertia with respect to the principal axes is zero.

B. *You Should be Able to:*

- Use the transformation equations for moments and products of inertia of areas.
- Determine the principal directions for an area.
- Determine the principal moments of inertia for an area.

C. Guided Practice Problems

S 33.1 For the shaded area shown, calculate the centroidal principal moments of inertia and determine the principal directions at the centroid.

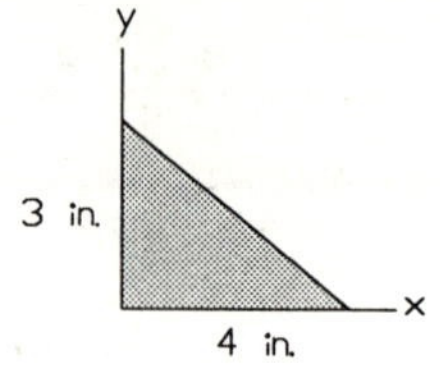

GUIDED SOLUTION

1. Calculate the centroidal inertial properties. Use Table 9.1 for right triangles to calculate $\bar{I}_x$, $\bar{I}_y$, and $\bar{I}_{xy}$.

✍

$\bar{I}_x =$

$\bar{I}_y =$

$\bar{I}_{xy} =$

2. Calculate the principal moments of inertia at the centroid. Use Eqs. (9.22) and (9.23) from the text:

✍ $$\bar{I}_1 = \frac{\bar{I}_x + \bar{I}_y}{2} + \sqrt{\left(\frac{\bar{I}_x - \bar{I}_y}{2}\right)^2 + \bar{I}_{xy}^2} =$$

$$\bar{I}_2 = \frac{\bar{I}_x + \bar{I}_y}{2} - \sqrt{\left(\frac{\bar{I}_x - \bar{I}_y}{2}\right)^2 + \bar{I}_{xy}^2} =$$

3. Determine the principal directions. Use the centroidal inertial properties in Eq. (9.20) from the text.

✍ $$\tan 2\theta = -\frac{2\bar{I}_{xy}}{\bar{I}_x - \bar{I}_y} =$$

$2\theta =$

$\theta_1 =$

$\theta_2 = \theta_1 + 90° =$

$\bar{I}_1 = 6.48 \text{ in.}^4$, $\bar{I}_2 = 1.85 \text{ in.}^4$
$\theta_1 = -29.9°$, $\theta_2 = 60.1°$

SOLUTIONS TO GUIDED PRACTICE PROBLEMS

S 1.1

1. $\beta = 180^\circ - 40^\circ = 140^\circ$

2. $R = \sqrt{P^2 + Q^2 - 2PQ\cos\beta} = \sqrt{150^2 + 75^2 - 2(150)(75)\cos 140^\circ}$
$= 213.0$ N

3. (a)

$$\frac{R}{\sin\beta} = \frac{P}{\sin\alpha}$$

$$\sin\alpha = \frac{P\sin\beta}{R} = \frac{150(0.6428)}{213} = 0.4527$$

$$\therefore \alpha = 26.9^\circ$$

(b) $\alpha + 30^\circ = 26.9^\circ + 30^\circ = 56.9^\circ$ ♦

S 2.1

Part (a)

1. $F_z = F\sin 60^\circ = 500(0.866) = 433.0$ N
$F_{xy} = F\cos 60^\circ = 500(0.50) = 250$ N

2. $F_x = F_{xy}\cos 50^\circ = 250(0.6428) = 160.7$ N
$F_y = F_{xy}\sin 50^\circ = 250(0.766) = 191.5$ N

3. $\mathbf{F} = F_x\mathbf{i} + F_y\mathbf{j} + F_z\mathbf{k} = 161\vec{i} + 192\vec{j} + 433\vec{k}$ N ♦

Part (b)

$$\cos\theta_x = \frac{F_x}{F} = \frac{160.7}{500} = 0.3214$$ ♦

$$\cos\theta_y = \frac{F_y}{F} = \frac{191.5}{500} = 0.3830$$ ♦

$$\cos\theta_z = \frac{F_z}{F} = \frac{433}{500} = 0.8660$$ ♦

Part (c)

$$\theta_x = \cos^{-1}\frac{F_x}{F} = \cos^{-1}(0.3214) = 71.3^\circ$$ ♦

$$\theta_y = \cos^{-1}\frac{F_y}{F} = \cos^{-1}(0.3830) = 67.5^\circ$$ ♦

$$\theta_z = \cos^{-1}\frac{F_z}{F} = \cos^{-1}(0.8660) = 30^\circ \text{ (also by inspection)}$$ ♦

S 2.2

Part (a)

1. $\overrightarrow{AB} = (0-4)\vec{i} + (0-12)\vec{j} + (3-0)\vec{k} = -4\vec{i} - 12\vec{j} + 3\vec{k}$ ft

2. $|\overrightarrow{AB}| = \sqrt{4^2 + 12^2 + 3^2} = 13.0$ ft

3. $\lambda_{AB} = \dfrac{\overrightarrow{AB}}{|\overrightarrow{AB}|} = \dfrac{-4\vec{i} - 12\vec{j} + 3\vec{k}}{13.0} = -0.3077\vec{i} - 0.9231\vec{j} + 0.2308\vec{k}$ ♦

Part (b)

$\mathbf{P} = P\lambda_{AB} = 130(-0.3077\vec{i} - 0.9231\vec{j} + 0.2308\vec{k}) = -40.0\vec{i} - 120\vec{j} + 30.0\vec{k}$ lb ♦

Part (c)

1. $\overrightarrow{AC} = -4\vec{i} + 3\vec{k}$ ft

2. $|\overrightarrow{AC}| = \sqrt{4^2 + 3^2} = 5$ ft

3. $\lambda_{AC} = \dfrac{\overrightarrow{AC}}{|\overrightarrow{AC}|} = \dfrac{-4\vec{i} + 3\vec{k}}{5} = -0.80\vec{i} + 0.60\vec{k}$

4. $\mathbf{Q} = Q\lambda_{AC} = 50(-0.80\vec{i} + 0.60\vec{k}) = -40.0\vec{i} + 30.0\vec{k}$ lb ♦

Part (d)

1. $\mathbf{R} = \mathbf{P} + \mathbf{Q} = (P_x + Q_x)\mathbf{i} + (P_y + Q_y)\mathbf{j} + (P_z + Q_z)\mathbf{k}$
$= (-40 - 40)\vec{i} + (-120\vec{j}) + (30 + 30)\vec{k} = -80.0\vec{i} - 120\vec{j} + 60.0\vec{k}$ lb ♦

2. $|\mathbf{R}| = \sqrt{R_x^2 + R_y^2 + R_z^2} = \sqrt{80^2 + 120^2 + 60^2} = 156$ lb ♦

S 3.1

Part (a)

1. (a) $\overrightarrow{AB} = 0.70\vec{i} + 0.40\vec{j} - 0.30\vec{k}$ m

(b) $\lambda_{AB} = \dfrac{\overrightarrow{AB}}{|\overrightarrow{AB}|} = \dfrac{0.70\vec{i} + 0.40\vec{j} - 0.30\vec{k}}{\sqrt{0.70^2 + 0.40^2 + 0.30^2}} = 0.8137\vec{i} + 0.4650\vec{j} - 0.3487\vec{k}$

(c) $\mathbf{P} = P\lambda_{AB} = 860\lambda_{AB} = 860(0.8137\vec{i} + 0.4650\vec{j} - 0.3487\vec{k}) = 700\vec{i} + 400\vec{j} - 300\vec{k}$ N

2. (a) $\overrightarrow{AC} = 0.40\vec{j} - 0.30\vec{k}$ m

(b) $\lambda_{AC} = \dfrac{\overrightarrow{AC}}{|\overrightarrow{AC}|} = \dfrac{0.40\vec{j} - 0.30\vec{k}}{\sqrt{0.40^2 + 0.30^2}} = 0.80\vec{j} - 0.60\vec{k}$

3. $P\cos\theta = \mathbf{P}\bullet\lambda_{AC} = 700(0) + 400(0.80) - 300(-0.6) = 500$ N ♦

Part (b)

$\theta = \cos^{-1}(\lambda_{AB}\bullet\lambda_{AC}) = \cos^{-1}[0.8137(0) + 0.4650(0.80) - 0.3487(-0.60)]$ ♦

$= \cos^{-1}(0.5812) = 54.5^\circ$

Part (c)

1. $\mathbf{Q} = Q\lambda_{AC} = 500\lambda_{AC} = 500(0.80\vec{j} - 0.60\vec{k}) = 400\vec{j} - 300\vec{k}$ N
2.

$$\mathbf{R} = \mathbf{P}\times\mathbf{Q} = \begin{vmatrix} \mathbf{i} & \mathbf{j} & \mathbf{k} \\ P_x & P_y & P_z \\ Q_x & Q_y & Q_z \end{vmatrix} = \begin{vmatrix} \vec{i} & \vec{j} & \vec{k} \\ 700 & 400 & -300 \\ 0 & 400 & -300 \end{vmatrix} = \begin{array}{l} \vec{i}[400(-300) - (-300)(400)] \\ -\vec{j}[700(-300)] + \vec{k}[700(400)] \end{array}$$ ♦

$= 210000\vec{j} + 280000\vec{k}\ N^2 = 210\vec{j} + 280\vec{k}\ kN^2$

S 3.2

1. $\mathbf{A} = A_y\mathbf{j} + A_z\mathbf{k} = (12/13)\,65\mathbf{j} + (5/13)\,65\mathbf{k} = 60.0\vec{j} + 25.0\vec{k}$ lb

   $\mathbf{B} = -B\mathbf{k} = -25.0\vec{k}$ lb

   $\mathbf{C} = -C_x\mathbf{i} + C_z\mathbf{k} = -(12/13)\,65\mathbf{i} + (5/13)\,65\mathbf{k} = -60.0\vec{i} + 25.0\vec{k}$ lb

2.

$$\mathbf{A}\times\mathbf{B}\bullet\mathbf{C} = \begin{vmatrix} A_x & A_y & A_z \\ B_x & B_y & B_z \\ C_x & C_y & C_z \end{vmatrix} = \begin{vmatrix} 0 & 60 & 25 \\ 0 & 0 & -25 \\ -60 & 0 & 25 \end{vmatrix} = -60[-(-25)(-60)] + 25(0)$$ ♦

$= 90000\ lb^3 = 90.0\times 10^3\ lb^3$

S 4.1

1. (a) $\overrightarrow{OA} = 8\vec{j} + 6\vec{k}$ m

   (b) $\lambda_{OA} = \dfrac{\overrightarrow{OA}}{|\overrightarrow{OA}|} = \dfrac{8\vec{j} + 6\vec{k}}{\sqrt{8^2 + 6^2}} = 0.80\vec{j} + 0.60\vec{k}$

   (c) $\mathbf{F}_1 = F_1\lambda_{OA} = 250\lambda_{OA} = 250(0.80\vec{j} + 0.60\vec{k}) = 200\vec{j} + 150\vec{k}$ kN

2. (a) $\lambda_{OB} = \vec{j}$

(b) $\mathbf{F}_2 = F_2\lambda_{OB} = 100\lambda_{OB} = 100\vec{j}$ kN

3. (a) $\overrightarrow{OC} = 4\vec{i} + 8\vec{j}$ m

(b) $\lambda_{OC} = \dfrac{\overrightarrow{OC}}{|\overrightarrow{OC}|} = \dfrac{4\vec{i}+8\vec{j}}{\sqrt{4^2+8^2}} = \dfrac{4\vec{i}+8\vec{j}}{8.944}$

(c) $\mathbf{F}_3 = F_3\lambda_{OC} = 200\lambda_{OC} = 200\left(\dfrac{4\vec{i}+8\vec{j}}{8.944}\right) = 89.44\vec{i} + 178.9\vec{j}$ kN

4. $\mathbf{R} = \mathbf{F}_1 + \mathbf{F}_2 + \mathbf{F}_3 = 89.44\vec{i} + (200+100+178.9)\vec{j} + 150\vec{k}$ ♦

$= 89.4\vec{i} + 479\vec{j} + 150\vec{k}$ kN

S 4.2

1. $\xrightarrow{+}$ $R_x = \Sigma F_x = 75\cos 30° + 90\cos 45° - 60\cos 60° - 100\cos 30°$

$= 64.95 + 63.64 - 30 - 86.6 = 12.0$ N

2. $+\uparrow$ $R_y = \Sigma F_y = 75\sin 30° - 90\sin 45° - 60\sin 60° + 100\sin 30°$

$= 37.5 - 63.64 - 51.96 + 50 = -28.1$ N

3. $\mathbf{R} = R_x\mathbf{i} + R_y\mathbf{j} = 12.0\vec{i} - 28.1\vec{j}$ N ♦

S 5.1

1. (a) $\overrightarrow{AB} = 4\vec{i} + 6\vec{j} - 3\vec{k}$ ft

(b) $\lambda_{AB} = \dfrac{\overrightarrow{AB}}{|\overrightarrow{AB}|} = \dfrac{4\vec{i}+6\vec{j}-3\vec{k}}{\sqrt{4^2+6^2+3^2}} = \dfrac{4\vec{i}+6\vec{j}-3\vec{k}}{7.810}$

$= 0.5121\vec{i} + 0.7682\vec{j} - 0.3841\vec{k}$

(c) $\mathbf{F} = F\lambda_{AB} = 500\lambda_{AB} = 500(0.5121\vec{i} + 0.7682\vec{j} - 0.3841\vec{k})$

$= 256.1\vec{i} + 384.1\vec{j} - 192.1\vec{k}$ lb

2. (a)

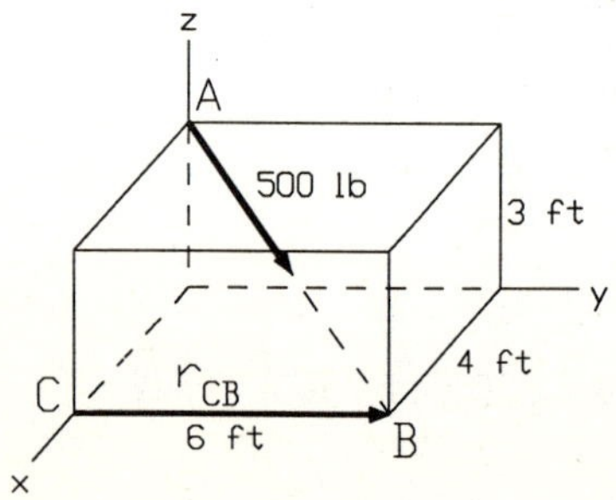

(b) $\mathbf{r}_{CB} = 6.0\vec{j}$ ft

3. $\mathbf{M}_C = \mathbf{r}_{CB} \times \mathbf{F}$

$$= \begin{vmatrix} \mathbf{i} & \mathbf{j} & \mathbf{k} \\ 0 & 6.0 & 0 \\ 256.1 & 384.1 & -192.1 \end{vmatrix} = -1152\vec{i} - 1536\vec{k} \text{ lb.ft}$$

4. $|\mathbf{M}_C| = \sqrt{M_{Cx}^{\ 2} + M_{Cy}^{\ 2} + M_{Cz}^{\ 2}} = \sqrt{1152^2 + 1536^2} = 1920$ lb.ft ♦

S 5.2

Part (a)

1. $F_y = (5/13)F = 5/13\,(390) = 150$ N ↑
   $F_x = (12/13)F = 12/13\,(390) = 360$ N ←

2.

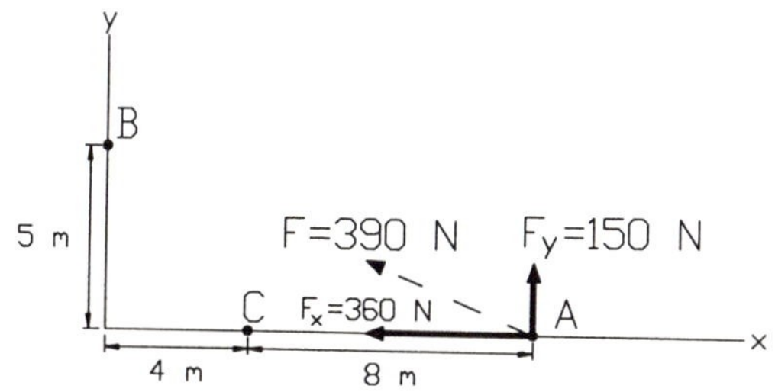

3. d for F_y = 8.0 m
   d for F_x = 0 (since the line of F_x passes through point C)

4. ↺+ $M_C = \Sigma(Fd) = 150(8.0) = 1200$ N·m CCW ♦

Part (b)

1.

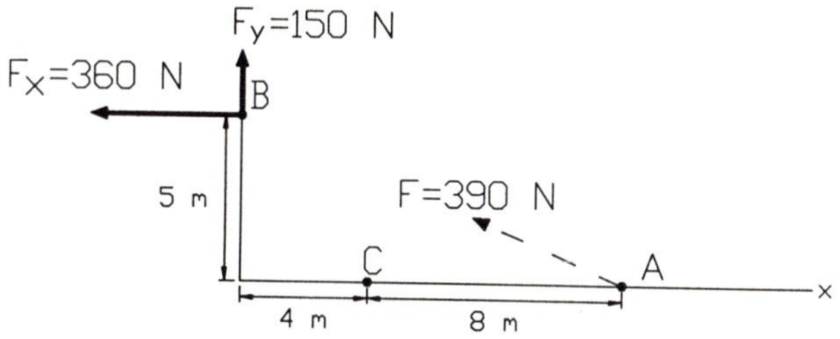

2. d for F_y = 4.0 m
   d for F_x = 5.0 m

3. $\circlearrowleft +$ $M_C = \Sigma(Fd) = -150(4.0) + 360(5.0) = 1200$ N·m CCW ◆

Part (c)

1. $d = 8(5/13) = 3.077$ m
2. $M_C = Fd = 390(3.077) = 1200$ N·m CCW ◆

S 6.1

1. (a) $\overrightarrow{AB} = 0.90\vec{i} + 1.20\vec{j} - 0.80\vec{k}$ m

   (b) $\lambda_{AB} = \dfrac{\overrightarrow{AB}}{|\overrightarrow{AB}|} = \dfrac{0.90\vec{i} + 1.20\vec{j} - 0.80\vec{k}}{\sqrt{0.90^2 + 1.20^2 + 0.80^2}} = \dfrac{0.90\vec{i} + 1.20\vec{j} - 0.80\vec{k}}{1.70}$

   (c) $\mathbf{P} = P\lambda_{AB} = 170\lambda_{AB} = 170\left(\dfrac{0.90\vec{i} + 1.20\vec{j} - 0.80\vec{k}}{1.70}\right) = 90.0\vec{i} + 120.0\vec{j} - 80.0\vec{k}$ kN

2. (a)

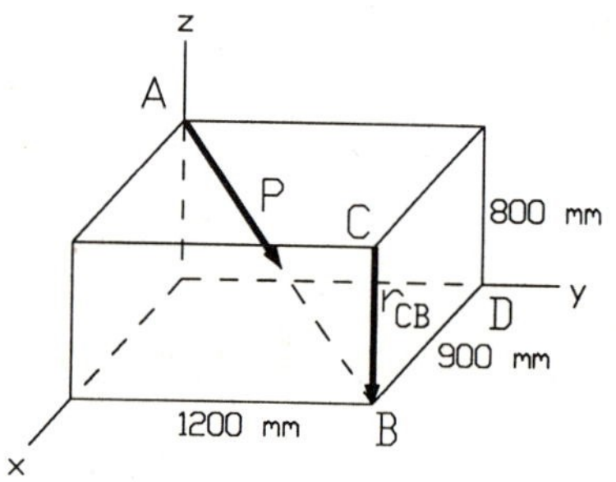

   (b) $\mathbf{r}_{CB} = -0.80\vec{k}$ m

3. $\lambda_{CD} = \dfrac{\overrightarrow{CD}}{|\overrightarrow{CD}|} = \dfrac{-0.90\vec{i} - 0.80\vec{k}}{\sqrt{0.90^2 + 0.80^2}} = \dfrac{-0.90\vec{i} - 0.80\vec{k}}{1.204} = -0.7475\vec{i} - 0.6645\vec{k}$

4. $|\mathbf{M}_{CD}| = (\mathbf{r} \times \mathbf{P}) \cdot \lambda_{CD} = \begin{vmatrix} 0 & 0 & -0.80 \\ 90.0 & 120.0 & -80.0 \\ -0.7475 & 0 & -0.6645 \end{vmatrix} = -0.80(120.0)(0.7475) = -71.76$ kN·m

5. $\mathbf{M}_{CD} = |\mathbf{M}_{CD}|\lambda_{CD} = -71.76(-0.7475\vec{i} - 0.6645\vec{k}) = 53.6\vec{i} + 47.7\vec{k}$ kN·m ◆

S 6.2

1. (a) $\overrightarrow{AB} = 4\vec{i} + 12\vec{j} - 3\vec{k}$ ft

(b) $\lambda_{AB} = \frac{\overrightarrow{AB}}{|\overrightarrow{AB}|} = \frac{4\vec{i}+12\vec{j}-3\vec{k}}{\sqrt{4^2+12^2+3^2}} = \frac{4\vec{i}+12\vec{j}-3\vec{k}}{13}$

(c) $\mathbf{F} = F\lambda_{AB} = 390\lambda_{AB} = 390\left(\frac{4\vec{i}+12\vec{j}-3\vec{k}}{13}\right) = 120\vec{i}+360\vec{j}-90\vec{k}$ lb

2. (a) $F_x = 120$ lb $\qquad F_y = 360$ lb $\qquad F_z = -90$ lb

(b)

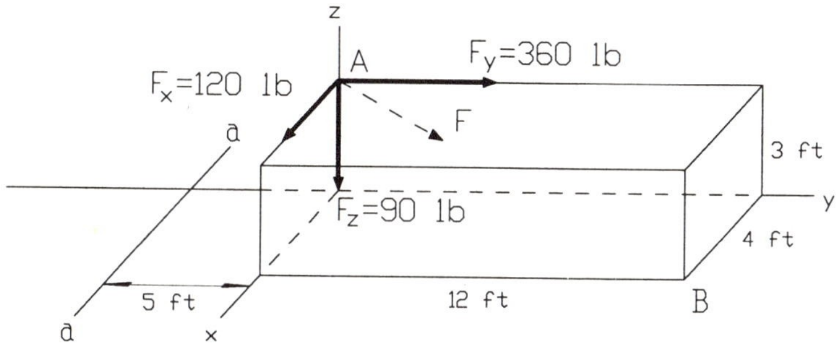

3. (a) F_x has no moment about the x-axis because it is parallel to the x axis.
F_y has a moment about the x-axis. Its moment arm is d = 3.0 ft
F_z has no moment about the x-axis because it intersects the x axis.

(b) $M_x = \Sigma(Fd) = -(F_y d) = -360(3.0) = -1080$ lb·ft ◆
(Since the turning effect of F_y about the x axis is clockwise, the vector M_x is in the negative x direction.)

4. (a) F_x has a moment about the y-axis. Its moment arm is d = 3.0 ft
F_y has no moment about the y-axis because it is parallel to the y axis.
F_z has no moment about the y-axis because it intersects the y axis.

(b) $M_y = \Sigma(Fd) = F_x d = 120(3.0) = 360$ lb·ft ◆

5. (a) F_x has no moment about the z-axis because it intersects the z axis.
F_y has no moment about the z-axis because it intersects the z axis
F_z has no moment about the z-axis because it is parallel (coincident) to the z axis.

(b) $M_z = \Sigma(Fd) = 0$ ◆

6. (a) F_x has no moment about the a-a axis because it is parallel to the a-a axis.
F_y has a moment about the a-a axis. Its moment arm is d_1 = 3.0 ft
F_z has a moment about the a-a axis. Its moment arm is d_2 = 5.0 ft

(b) $\overset{a}{\curvearrowleft +}\ M_{a\text{-}a} = \Sigma(Fd) = -(F_y d_1) - (F_z d_2) = -360(3.0) - 90(5.0)$ ◆
$= -1530$ lb·ft

S 7.1

Part (a)

1. (a)

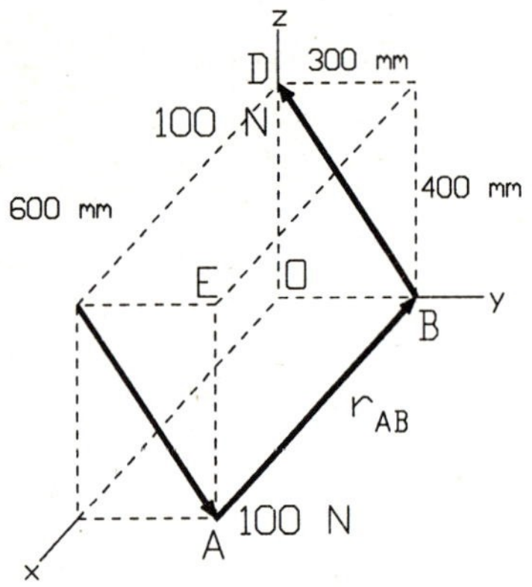

(b) $\mathbf{r}_{AB} = -0.60\vec{i}$ m

2. (a) $\lambda_{BD} = \dfrac{\vec{BD}}{|\vec{BD}|} = \dfrac{-0.30\vec{j}+0.40\vec{k}}{\sqrt{0.30^2+0.40^2}} = -0.60\vec{j} + 0.80\vec{k}$

(b) $\mathbf{F} = 100\lambda_{BD} = 100(-0.60\vec{j}+0.80\vec{k}) = -60.0\vec{j} + 80.0\vec{k}$ N

3.

$$\mathbf{C} = \mathbf{M}_A = \mathbf{r}_{AB} \times \mathbf{F} = \begin{vmatrix} \mathbf{i} & \mathbf{j} & \mathbf{k} \\ -0.60 & 0 & 0 \\ 0 & -60.0 & 80.0 \end{vmatrix} = -\vec{j}[-0.60(80.0)] + \vec{k}[-0.60(-60.0)]$$ ◆

$= 48.0\vec{j} + 36.0\vec{k}$ N·m

Part (b)

$\mathbf{M}_O = \mathbf{M}_A = \mathbf{C} = 48.0\vec{j} + 36.0\vec{k}$ N·m ◆

Part (c)

1. $\lambda_{OE} = \dfrac{\vec{OE}}{|\vec{OE}|} = \dfrac{0.60\vec{i}+0.30\vec{j}+0.40\vec{k}}{\sqrt{0.60^2+0.30^2+0.40^2}} = \dfrac{0.60\vec{i}+0.30\vec{j}+0.40\vec{k}}{0.781} = 0.7682\vec{i}+0.3841\vec{j}+0.5121\vec{k}$

2. (a) $M_{OE} = \mathbf{C} \cdot \lambda_{OE} = 48.0(0.3841) + 36.0(0.5121) = 36.87$ N·m

(b) $\mathbf{M}_{OE} = M_{OE}\lambda_{OE} = 36.87(0.7682\vec{i} + 0.3841\vec{j} + 0.5121\vec{k})$ ♦
$= 28.3\vec{i} + 14.2\vec{j} + 18.9\vec{k}$ N·m

S 8.1

1. (a) $\lambda_{AB} = \dfrac{\overrightarrow{AB}}{|\overrightarrow{AB}|} = \dfrac{-8.0\vec{i} - 8.0\vec{j} + 4.0\vec{k}}{\sqrt{8.0^2 + 8.0^2 + 4.0^2}} = \dfrac{-8.0\vec{i} - 8.0\vec{j} + 4.0\vec{k}}{12.0}$

(b) $\mathbf{F} = 60\lambda_{AB} = 60\left(\dfrac{-8.0\vec{i} - 8.0\vec{j} + 4.0\vec{k}}{12.0}\right) = -40.0\vec{i} - 40.0\vec{j} + 20.0\vec{k}$ lb

2. (a) $\mathbf{r}_{DB} = -3.0\vec{i} + 2.0\vec{k}$ in.

(b)

$$\mathbf{C}^T = \mathbf{M}_D = \mathbf{r}_{DB} \times \mathbf{F} = \begin{vmatrix} \mathbf{i} & \mathbf{j} & \mathbf{k} \\ -3.0 & 0 & 2.0 \\ -40.0 & -40.0 & 20.0 \end{vmatrix} = \vec{i}[-2.0(-40.0)] - \vec{j}[-3.0(20.0) - 2.0(-40.0)] + \vec{k}[-3.0(-40.0)]$$

$= 80.0\vec{i} - 20.0\vec{j} + 120\vec{k}$ lb·in.

3. (a) $\lambda_{AE} = \dfrac{\overrightarrow{AE}}{|\overrightarrow{AE}|} = \dfrac{-8.0\vec{i} + 1.0\vec{j} + 4.0\vec{k}}{\sqrt{8.0^2 + 1.0^2 + 4.0^2}} = \dfrac{-8.0\vec{i} + \vec{j} + 4.0\vec{k}}{9.0}$

(b) $\mathbf{C} = 180\lambda_{AE} = 180\left(\dfrac{-8.0\vec{i} + \vec{j} + 4.0\vec{k}}{9.0}\right) = -160\vec{i} + 20.0\vec{j} + 80.0\vec{k}$ lb·in.

4. $\mathbf{C}^R = \mathbf{C} + \mathbf{C}^T = (-160 + 80.0)\vec{i} + (20.0 - 20.0)\vec{j} + (80.0 + 120)\vec{k}$
$= -80.0\vec{i} + 200\vec{k}$ lb·in.

5. $\mathbf{F} = -40.0\vec{i} - 40.0\vec{j} + 20.0\vec{k}$ lb ♦
$\mathbf{C}^R = -80.0\vec{i} + 200\vec{k}$ lb·in. ♦

S 9.1

1. (a) $\mathbf{F}_1 = 200\vec{k}$ N
$\mathbf{F}_2 = 150\vec{i}$ N
$\mathbf{F}_3 = 250\vec{j}$ N

(b) $\mathbf{R} = \Sigma\mathbf{F} = 150\vec{i} + 250\vec{j} + 200\vec{k}$ N

2.

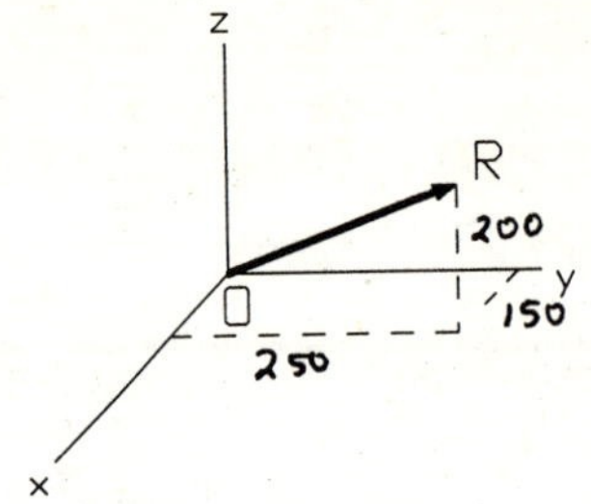

3. For $\mathbf{F}_1$: $\mathbf{M}_O = (200 \times 8)\vec{i} = 1600\vec{i}$ N·m

For $\mathbf{F}_2$: $\mathbf{M}_O = (150 \times 3)\vec{j} = 450\vec{j}$ N·m

For $\mathbf{F}_3$: $\mathbf{M}_O = (250 \times 5)\vec{k} = 1250\vec{k}$ N·m

For $\mathbf{C}$: $\mathbf{M}_O = \mathbf{C} = 500\vec{j}$ N·m

$\mathbf{C}^R = \Sigma\mathbf{M}_O = 1600\vec{i} + (450 + 500)\vec{j} + 1250\vec{k} = 1600\vec{i} + 950\vec{j} + 1250\vec{k}$ N·m

4. $\mathbf{R} = 150\vec{i} + 250\vec{j} + 200\vec{k}$ N ♦

$\mathbf{C}^R = 1600\vec{i} + 950\vec{j} + 1250\vec{k}$ N·m ♦

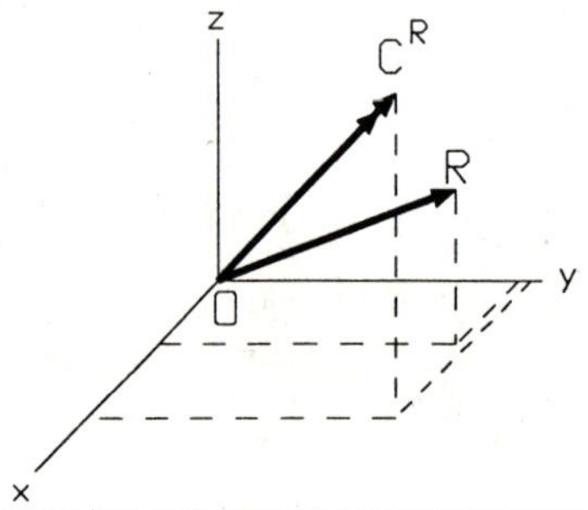

S 10.1

1. (a) $R_x = \Sigma F_x$: $\overset{+}{\rightarrow}$ $R_x = 80 + \frac{4}{5}(50) = 120$ N →

$R_y = \Sigma F_y$: $+\uparrow$ $R_y = 60 - \frac{3}{5}(50) = 30$ N ↑

(b) $R = \sqrt{R_x^2 + R_y^2} = \sqrt{120^2 + 30^2} = 124$ N

2. (a) $\circlearrowleft +$ $\Sigma M_O = -90 + 60(5) - 80(3) - \frac{3}{5}(50)(5) = -180$
$= 180$ N·m ↻

(b) $d = \Sigma M_O / R = 180/124 = 1.45$ m

3.

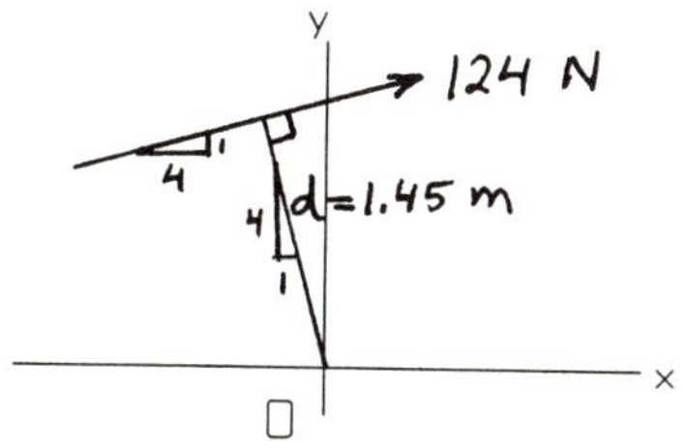

♦

S 10.2

1\. (a) $R_x = \Sigma F_x$: $\xrightarrow{+}$ $R_x = \frac{3}{5}(50) - 30 = 0$

$R_y = \Sigma F_y$: $+\uparrow$ $R_y = \frac{4}{5}(50) - 40 = 0$

(b) ↺+ $\Sigma M_O = 30(3) + \frac{4}{5}(50)(5) = 90 + 200 = 290$ lb·in ↺

2\.

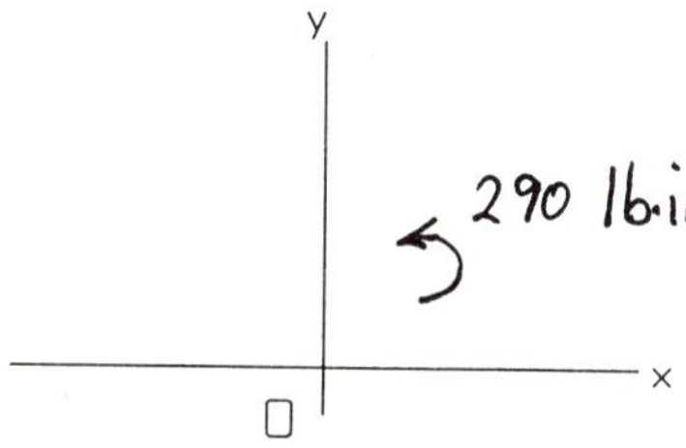

Since the moment of a couple is the same about any point, couples can be shown anywhere.

♦

S 11.1

1\. $+\uparrow$ $R = \Sigma F_z = 100 - 200 + 300 = 200$ lb ↑

2\. (x, +) $\Sigma M_x = 300(6) - 1000 = 800$ lb·ft

(+, y) $\Sigma M_y = 200(3) - 300(2) = 0$

3\. $\bar{x} = -\Sigma M_y / R = 0/200 = 0$

$\bar{y} = \Sigma M_x / R = 800/200 = 4.0$ ft

4\.

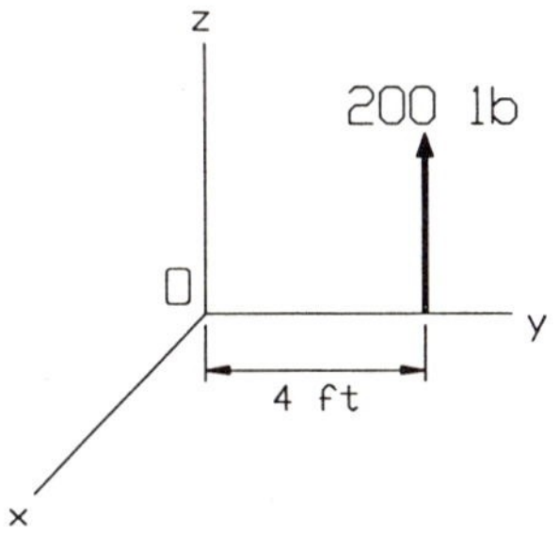

♦

S 11.2

1. $+\uparrow \; R = \Sigma F_z = 100 - 50 - 50 = 0$

   Since $\Sigma F_z = 0$, the resultant is not a force.

2. x + $\quad \Sigma M_x = 100(2) - 50(2) = 100 \text{ N·m}$

   y + $\quad \Sigma M_y = -100(4) = -400 \text{ N·m}$

3. $C^R = 100\vec{i} - 400\vec{j} \text{ N·m}$ ◆

S 11.3

1. (a) $\lambda = \mathbf{R}/R = \dfrac{300\vec{i} + 450\vec{j} + 200\vec{k}}{\sqrt{300^2 + 450^2 + 200^2}} = 0.5203\vec{i} + 0.7804\vec{j} + 0.3469\vec{k}$

   (b) $C^R{}_t = C^R \cdot \lambda = 500(0.5203) + 400(0.7804) + 600(0.3469) = 780.4 \text{ N·m}$

   (c) $C^R{}_t = C^R{}_t \lambda = 780.4(0.5203\vec{i} + 0.7804\vec{j} + 0.3469\vec{k})$ ◆

   $= 406\vec{i} + 609\vec{j} + 270.7\vec{k} \text{ N·m}$

2. $C^R{}_n = C^R - C^R{}_t = (500 - 406)\vec{i} + (400 - 609)\vec{j} + (600 - 270.7)\vec{k}$

   $= 94.0\vec{i} - 209.0\vec{j} + 329.3\vec{k} \text{ N·m}$

3. (a)

$$C^R{}_n = \begin{vmatrix} \mathbf{i} & \mathbf{j} & \mathbf{k} \\ x & y & 0 \\ 300 & 450 & 200 \end{vmatrix}$$

   (b) Expanding the determinant: $\vec{i}(200y) - \vec{j}(200x) + \vec{k}(450x - 300y)$

   (c) (**i** component): $200y = 94.0$

   $y = 94.0/200 = 0.470 \text{ m}$ ◆

   (**j** component): $-200x = -209$

   $x = 209/200 = 1.045 \text{ m}$ ◆

S 12.1

1. $P_1 = A_1 = 24(20) = 480 \text{ lb}$

   $P_2 = A_2 = \frac{1}{2}(24)(60 - 20) = 480 \text{ lb}$

2. $+\downarrow \; R = P_1 + P_2 = 480 + 480 = 960 \text{ lb}$ ◆

3. $\bar{x}_1 = 12.0 \text{ ft}$

   $\bar{x}_2 = \frac{1}{3}(24) = 8.0 \text{ ft}$

4.

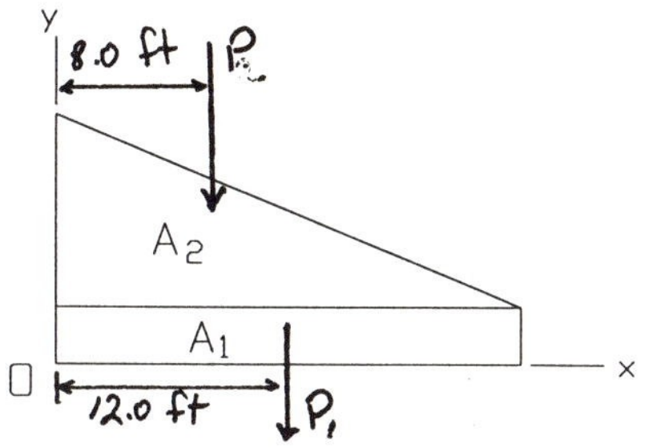

5. $+\circlearrowleft$ $\Sigma M_O = P_1 \bar{x}_1 + P_2 \bar{x}_2 = 480(12) + 480(8) = 9600$ lb·ft

$\bar{x} = \Sigma M_O / R = 9600/960 = 10.0$ ft ◆

S 12.2

1. $R = pA = 720\left[\frac{1}{2}(0.750)(0.750)\right] = 202.5$ ◆
   $= 203$ N
2. By symmetry: $\bar{x} = 750/2 = 375$ mm ◆
   From Table 3.1: $\bar{y} = \frac{2}{3}(750) = 500$ mm ◆

S 13.1

1, 2. The free-body diagram of bar AB:

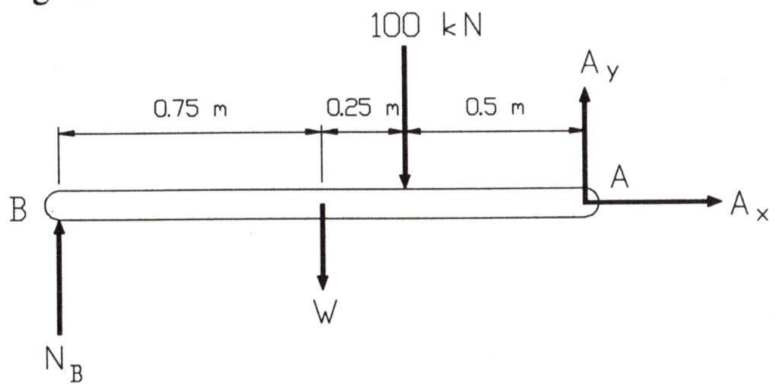

W = 20(9.81) = 196 N ◆

3. The unknowns are: A_x, A_y, N_B ◆
   Number of unkowns = 3 ◆

S 13.2

1, 2. The free-body diagram of bar BC:

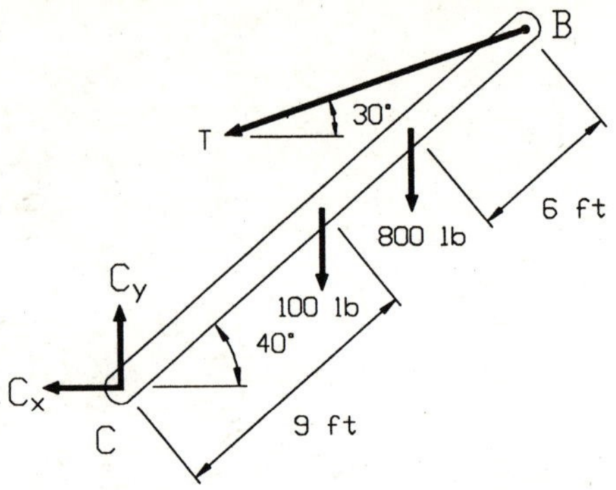

3. The unknowns are: C_x, C_y, T ♦

Number of unknowns = 3 ♦

S 14.1

1. $\overset{+}{\rightarrow}$ $\Sigma F_X = 0$: $A_x - 1000 - 375\cos 45° + 750\cos 60° = 0$ ♦

$A_x = 890$ lb →

2. ↺+ $\Sigma M_A = 0$: $250(7.5) + N_B(10) - 750\sin 60°(10) - 375\sin 45°(3) = 0$ ♦

$N_B = 542$ lb ↑

3. +↑ $\Sigma F_y = 0$: $A_y + 250 + N_B - 375\sin 45° - 750\sin 60° = 0$ ♦

$A_y = 123$ lb ↑

4. ↺+ $\Sigma M_B = 375\sin 45°(7) - 123(10) - 250(2.5) = 1856 - 1230 - 625 = 1$

≈ 0

S 15.1

1.

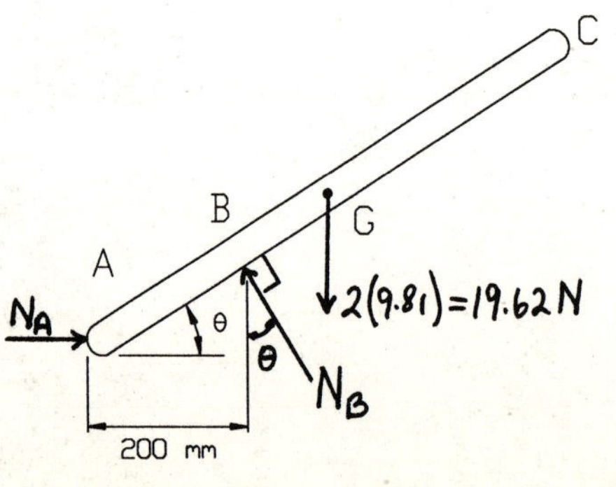

2. (a) $+\uparrow$ $\Sigma F_y = 0$: $N_B \cos\theta - 19.62 = 0$

(b) $\circlearrowleft +$ $\Sigma M_A = 0$: $N_B(200/\cos\theta) - 19.62(350\cos\theta) = 0$

(c) $\xrightarrow{+}$ $\Sigma F_x = 0$: $N_A - N_B \sin\theta = 0$

3. (a) Equations from 2(a) and (b): ◆

2(a) $N_B = 19.62/\cos\theta$

2(b) $N_B = \dfrac{19.62(350\cos\theta)\cos\theta}{200} = 34.34\cos^2\theta$

$\dfrac{19.62}{\cos\theta} = 34.34\cos^2\theta$

$\cos^3\theta = 0.571$

$\theta = 33.9^\circ$

$N_B = 19.62/\cos 33.9 = 23.6$ N

(b) Equation from 2(c): $N_A = 23.6 \sin 33.9 = 13.2$ N ◆

S 16.1

Part (a)

1.

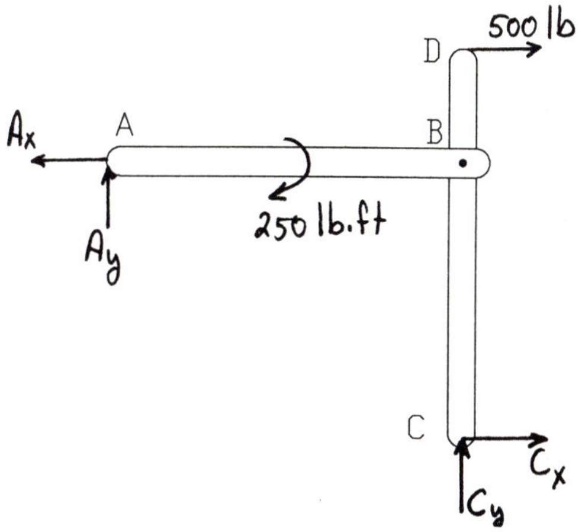

2. (a) Number of independent equilibrium equations = 3 ◆

(b) The number of unknowns on the FBD = 4 : A_x, A_y, C_x, C_y ◆

Part (b)

1.

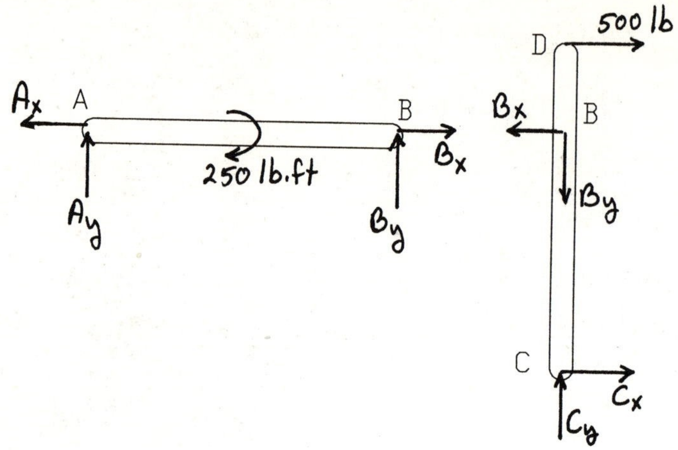

2. (a) Number of independent equilibrium equations = 6 ♦

(b) The number of unknowns on the FBDs = 6 : $A_x, A_y, B_x, B_y, C_x, C_y$ ♦

S 16.2

Part (a)

1.

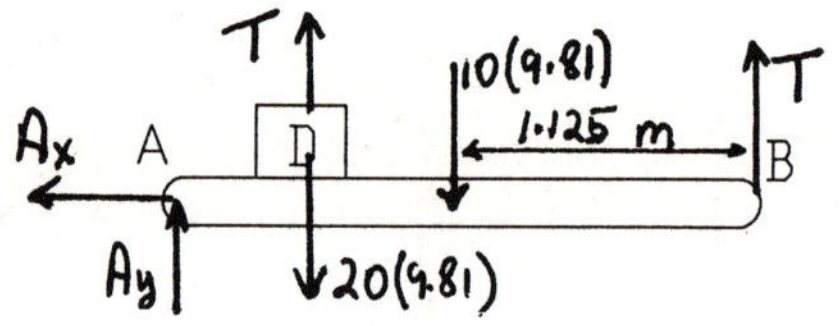

2. (a) Number of independent equilibrium equations = 3 ♦

(b) The number of unknowns on the FBD = 3 : A_x, A_y, T ♦

Part (b)

1.

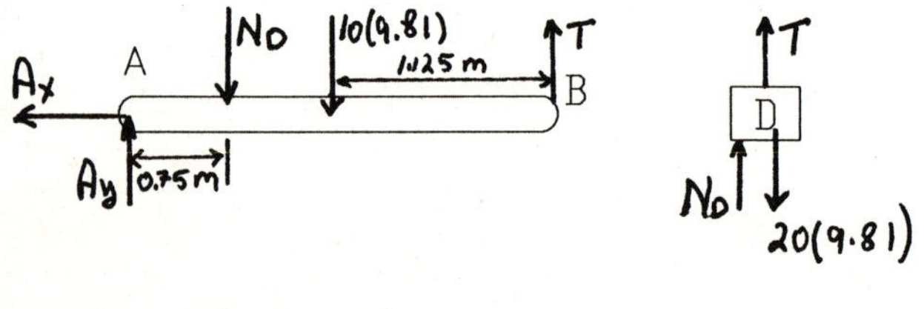

2. (a) Number of independent equilibrium equations = 4 ♦

(b) The number of unknowns on the FBDs = 4 : A_x, A_y, N_D, T ♦

S 17.1

1.

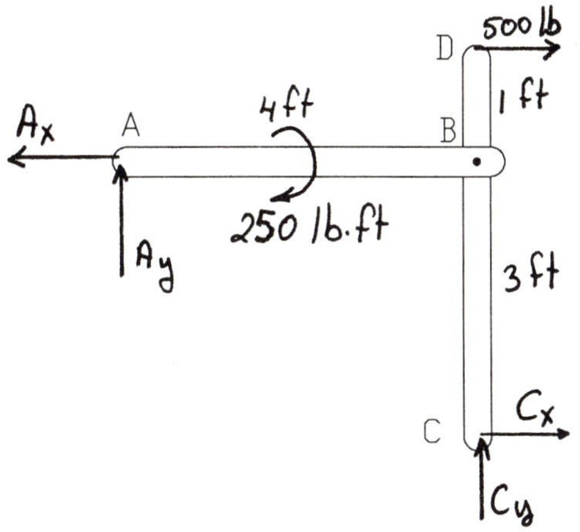

2.

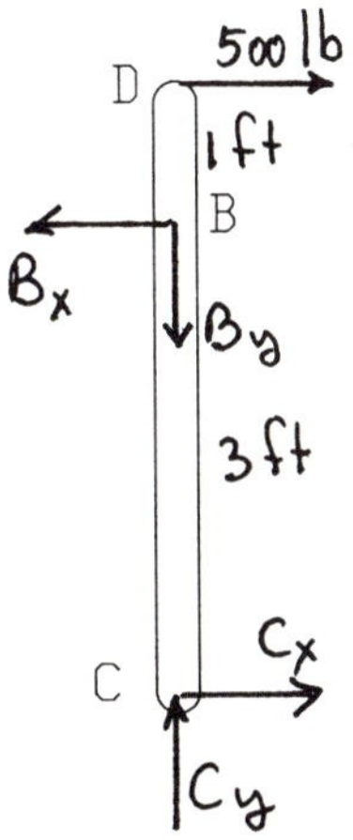

3. (a) The total number of independent equilibrium equations = 6

   (b) The number of unknowns on the FBDs = 6 : $A_x, A_y, C_x, C_y, B_x, B_y$

4. (a) $\circlearrowleft +$ $\Sigma M_A = 0$: $-250 - 500(1) + C_y(4) + C_x(3) = 0$

   $C_y = 187.5 - 0.75C_x$

   (b) $\circlearrowleft +$ $\Sigma M_B = 0$: $C_x(3) - 500(1) = 0$

   $C_x = 166.7 \text{ lb}$

   (c) Solving Eqn. (1) and Eqn. (2) simultaneously:

   $C_y = 187.5 - 0.75(166.7) = 62.5 \text{ lb}$

5. $R_C = \sqrt{C_x^2 + C_y^2} = \sqrt{166.7^2 + 62.5^2} = 178 \text{ lb}$ ♦

   $\theta = \tan^{-1}(C_y / C_x) = \tan^{-1}(62.5/166.7) = 20.6°$ ♦

S 17.2

1.

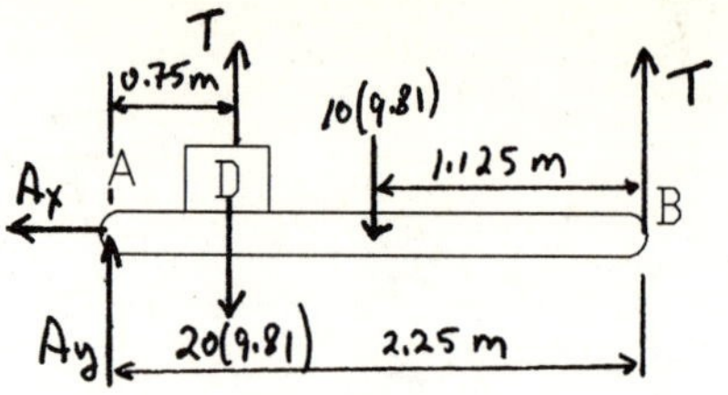

2.

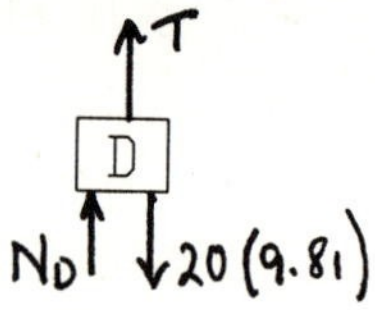

3. (a) ↺+ $\Sigma M_A = 0$: $T(0.75) + T(2.25) - 20(9.81)(0.75) - 10(9.81)(1.125) = 0$

$T = 85.8 \text{ N}$

(b) +↑ $\Sigma F_y = 0$: $85.8 + N_D - 20(9.81) = 0$

$N_D = 110 \text{ N}$

♦

S 17.3

1.

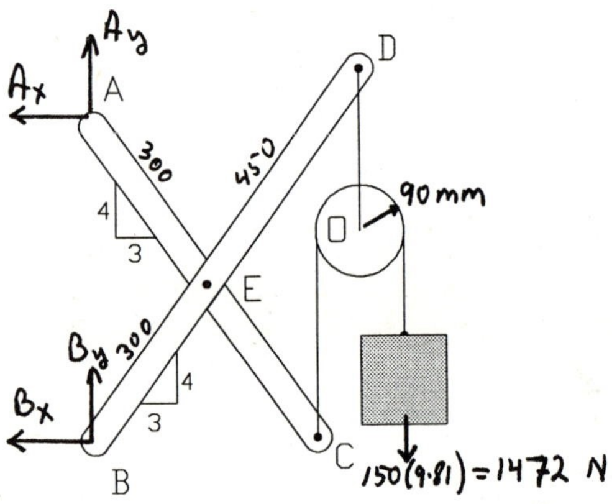

2. The number of unknowns on the FBD = 4 : A_x, A_y, B_x, B_y

Number of independent equilibrium equations = 3

3.

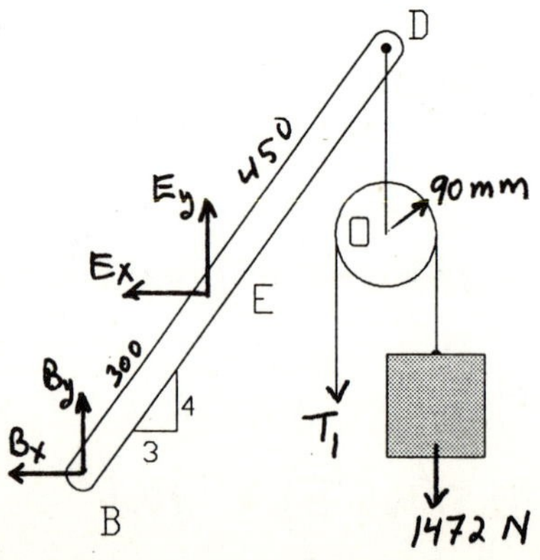

4. The number of unknowns on the FBDs = 7: $A_x, A_y, B_x, B_y, E_x, E_y, T_1$

Number of independent equilibrium equations = 6

5.

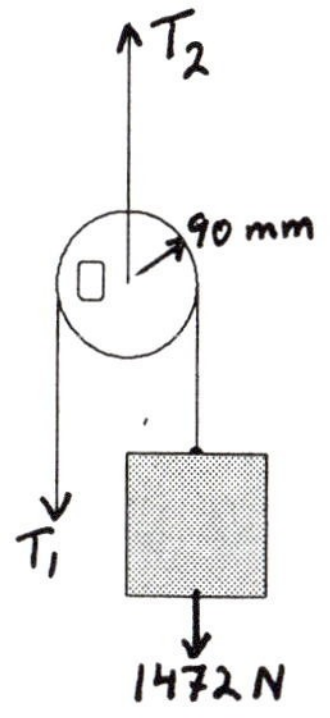

6. The number of unknowns on the FBDs = 8 : $A_x, A_y, B_x, B_y, E_x, E_y, T_1, T_2$

Number of independent equilibrium equations = 8

7. ↺+ $\Sigma M_A = 0$: $-B_x\left(\frac{4}{5}\right)(2)(300) - 1472\left[\frac{3}{5}(750) + 90\right] = 0$

$B_x = -1656$ N

↺+ $\Sigma M_O = 0$: $T_1(90) - 1472(90) = 0$

$T_1 = 1472$ N

↺+ $\Sigma M_E = 0$: $-B_y\left(\frac{3}{5}\right)(300) - (-1656)\left(\frac{4}{5}\right)(300) - 1472\left[\frac{3}{5}(450) - 90\right]$

$- 1472\left[\frac{3}{5}(450) + 90\right] = 0$

$B_y = 2208$ N

8. $R_B = \sqrt{B_x^2 + B_y^2} = \sqrt{1656^2 + 2208^2} = 2760$ N ◆

S 18.1

1.

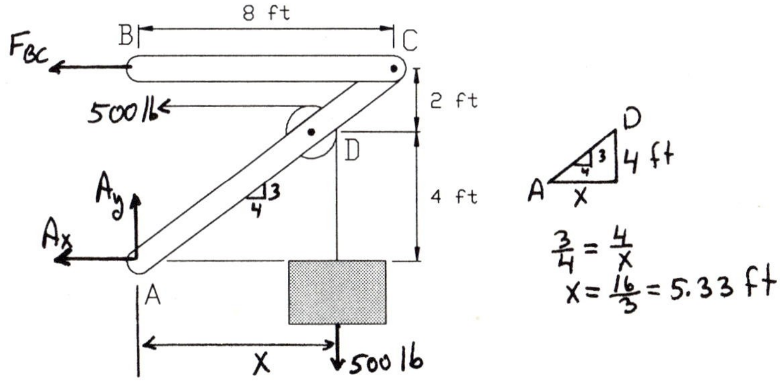

2. (a) The number of unknowns on the FBD of the entire frame = 3 : F_{BC}, A_x, A_y

(b) Total number of independent equilibrium equations = 3

3. ↺+ $\Sigma M_A = 0$: $F_{BC}(6) + 500(4) - 500(5.33) = 0$

$F_{BC} = 111$ lb ← ◆

S 18.2

1.

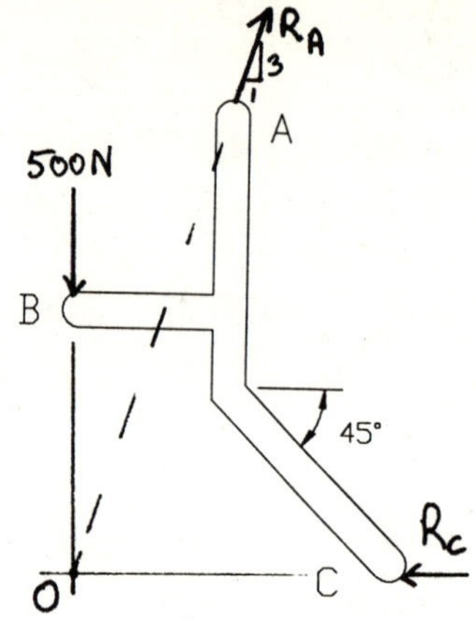

2. (a) The number of unknowns on the FBD of the entire frame = 2 : R_A, R_c
   (b) Total number of independent equilibrium equations = 2 (concurrent force system)

3. (a) $+\uparrow$ $\Sigma F_y = 0$: $3/\sqrt{10}\, R_A - 500 = 0$ ◆
   $R_A = 527$ N

   (b) $\xrightarrow{+}$ $\Sigma F_x = 0$: $1/\sqrt{10}\,(527) - R_c = 0$ ◆
   $R_c = 167$ N

S 19.1

1.

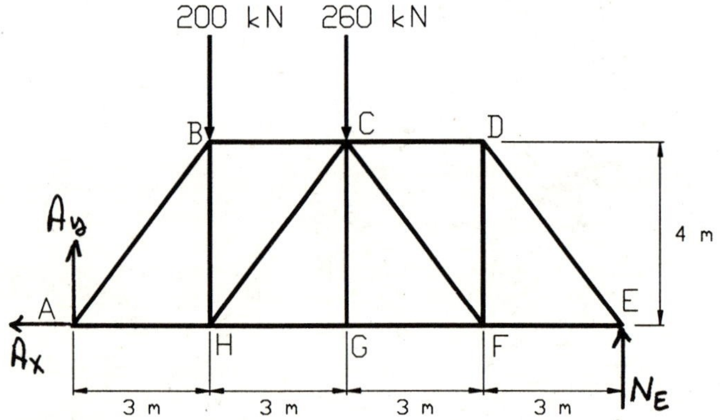

2. The number of unknowns on the FBD of the entire truss = 3 : A_x, A_y, N_E
   Number of available equilibrium equations = 3
   Can the external reactions be determined from this FBD? yes

3. (a) $\circlearrowleft +$ $\Sigma M_E = 0$: $-A_y(12) + 200(9) + 260(6) = 0$
   $A_y = 280$ kN ↑

   (b) $\xrightarrow{+}$ $\Sigma F_x = 0$: $A_x = 0$

4.

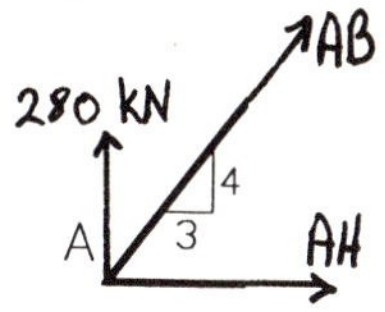

5. (a) $+\uparrow$ $\Sigma F_y = 0$: $280 + \frac{4}{5} AB = 0$

$AB = -350$ $AB = 350$ kN Compression ♦

(b) $\xrightarrow{+}$ $\Sigma F_x = 0$: $\frac{3}{5}(-350) + AH = 0$

$AH = 210$ kN Tension ♦

6.

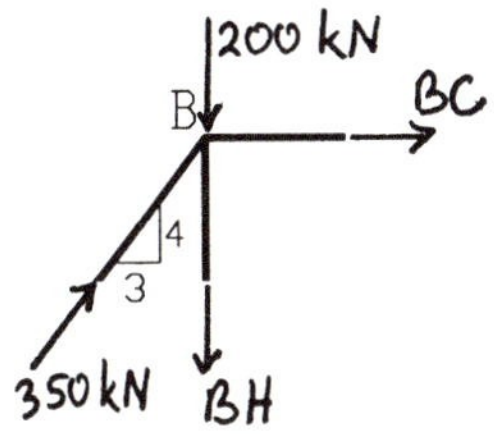

7. (a) $+\uparrow$ $\Sigma F_y = 0$: $\frac{4}{5}(350) - BH - 200 = 0$

$BH = 80$ kN Tension ♦

(b) $\xrightarrow{+}$ $\Sigma F_x = 0$: $\frac{3}{5}(350) + BC = 0$

$BC = -210$ $BC = 210$ kN Compression ♦

S 20.1

1, 2.

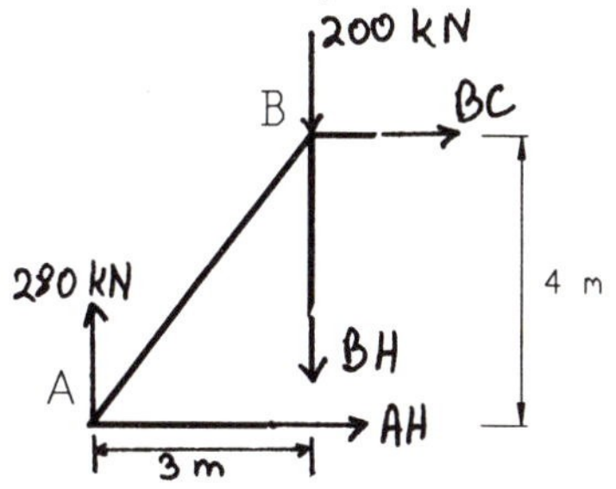

3. (a) $+\uparrow$ $\Sigma F_y = 0$: $280 - 200 - BH = 0$

$BH = 80$ kN Tension ♦

(b) $\circlearrowleft +$ $\Sigma M_A = 0$: $-80(3) - 200(3) - BC(4) = 0$

$BC = -210$ $BC = 210$ kN Compression ♦

S 21.1

1. The free-body diagram of the mast AD:

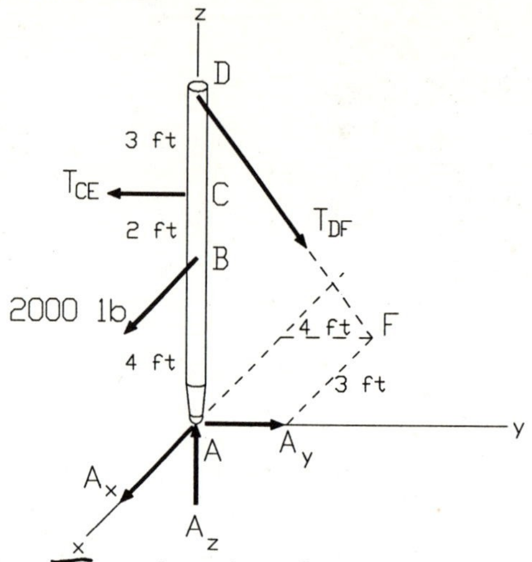

2. (a) The unknowns are: T_{DF}, T_{CE}, A_x, A_y, A_z ◆
   (b) The number of unknowns = 5 ◆

S 22.1

1. (a) $\mathbf{F} = 2000\vec{i}$ lb

   (b) $\mathbf{T}_{CE} = T_{CE}\lambda_{CE} = -T_{CE}\vec{j}$

   (c) $\mathbf{T}_{DF} = T_{DF}\lambda_{DF} = T_{DF}\left(\frac{-3\vec{i}+4\vec{j}-9\vec{k}}{10.3}\right) = T_{DF}\left(-0.2913\vec{i}+0.3883\vec{j}-0.8738\vec{k}\right)$

   (d) $\mathbf{r}_{AB} = 4\vec{k}$ ft

   (e) $\mathbf{r}_{AC} = 6\vec{k}$ ft

   (f) $\mathbf{r}_{AD} = 9\vec{k}$ ft

   (g) $\Sigma\mathbf{M}_A = \mathbf{0}$: $\Sigma(\mathbf{r}\times\mathbf{F}) = \mathbf{0}$

$$\frac{T_{DF}}{10.3}\begin{vmatrix} \mathbf{i} & \mathbf{j} & \mathbf{k} \\ 0 & 0 & 9 \\ -3 & 4 & -9 \end{vmatrix} + \begin{vmatrix} \mathbf{i} & \mathbf{j} & \mathbf{k} \\ 0 & 0 & 4 \\ 2000 & 0 & 0 \end{vmatrix} + \begin{vmatrix} \mathbf{i} & \mathbf{j} & \mathbf{k} \\ 0 & 0 & 6 \\ 0 & -T_{CE} & 0 \end{vmatrix} = \mathbf{0}$$

$$\frac{T_{DF}}{10.3}\left[(-9)(4)\vec{i} - \vec{j}(-9)(-3)\right] - \vec{j}(-4)(2000) + \vec{i}(-6)(-T_{CE}) = \vec{0}$$

$$-3.495T_{DF}\vec{i} - 2.621T_{DF}\vec{j} + 8000\vec{j} + 6T_{CE}\vec{i} = \vec{0}$$

(h) Σ(**i** components) = 0: $-3.495\,T_{DF} + 6\,T_{CE} = 0$

Σ(**j** components) = 0: $-2.621\,T_{DF} + 8000 = 0$

Solving for T_{DF} and T_{CE}: $T_{DF} = 3052$ lb ♦

$-3.495(3052) + 6T_{CE} = 0$

$T_{CE} = 1778$ lb

2. (a) $\Sigma F_x = 0$: $A_x + 2000 - 3052(0.2913) = 0$

$A_x = -1111$ lb

(b) $\Sigma F_y = 0$: $A_y - 1778 + 3052(0.3883) = 0$

$A_y = 593$ lb

(c) $\Sigma F_z = 0$: $A_z - 3052(0.8738) = 0$

$A_z = 2667$ lb

(d) $A = \sqrt{A_x^2 + A_y^2 + A_z^2} = \sqrt{1111^2 + 593^2 + 2667^2} = 2950$ lb ♦

S 23.1

1.

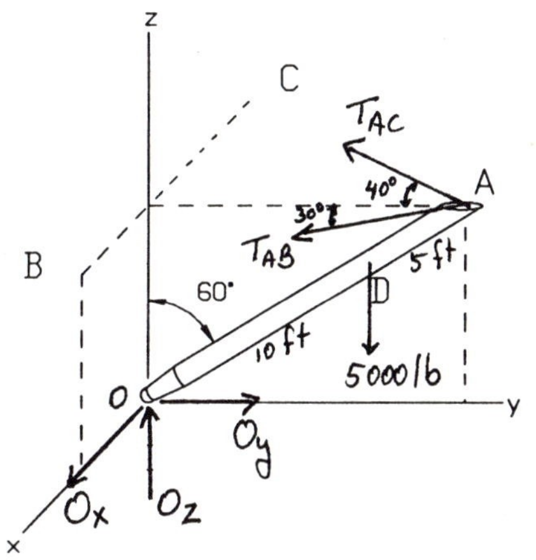

2. (a) The unknowns on the FBD: O_x, O_y, O_z, T_{AB}, T_{AC}

(b) Number of independent equilibrium equations = 5 (all forces intersect axis OA)

3. (a) $\mathbf{T}_{AC} = T_{AC}(-\sin 40°\vec{i} - \cos 40°\vec{j}) = T_{AC}(-0.6428\vec{i} - 0.7660\vec{j})$ lb

$\mathbf{T}_{AB} = T_{AB}(\sin 30°\vec{i} - \cos 30°\vec{j}) = T_{AB}(0.50\vec{i} - 0.8660\vec{j})$ lb

$\mathbf{P} = -5000\vec{k}$ lb

(b) $\mathbf{r}_{OA} = 15\sin 60°\vec{j} + 15\cos 60°\vec{k} = 12.99\vec{j} + 7.50\vec{k}$ ft

$\mathbf{r}_{OD} = 10\sin 60°\vec{j} + 10\cos 60°\vec{k} = 8.660\vec{j} + 5.0\vec{k}$ ft

(c)

$$T_{AC}\begin{vmatrix} \mathbf{i} & \mathbf{j} & \mathbf{k} \\ 0 & 12.99 & 7.50 \\ -0.6428 & -0.7660 & 0 \end{vmatrix} + T_{AB}\begin{vmatrix} \mathbf{i} & \mathbf{j} & \mathbf{k} \\ 0 & 12.99 & 7.50 \\ 0.50 & -0.8660 & 0 \end{vmatrix} + \begin{vmatrix} \mathbf{i} & \mathbf{j} & \mathbf{k} \\ 0 & 8.660 & 5.0 \\ 0 & 0 & -5000 \end{vmatrix} = \mathbf{0}$$

$$T_{AC}\vec{i}(-7.50)(-0.7660) - T_{AC}\vec{j}(-7.50)(-0.6428) + T_{AC}\vec{k}(-12.99)(-0.6428) + T_{AB}\vec{i}(-7.50)(-0.866)$$
$$- T_{AB}\vec{j}(-7.50)(0.50) + T_{AB}\vec{k}(-12.99)(0.50) + \vec{i}(8.660)(-5000) = \vec{0}$$

$$5.745\,T_{AC}\vec{i} + 6.495\,T_{AB}\vec{i} - 43300\,\vec{i} - 4.817\,T_{AC}\vec{j} + 3.750\,T_{AB}\vec{j} + 8.350\,T_{AC}\vec{k} - 6.50\,T_{AB}\vec{k} = \vec{0}$$

Σ(**i** components) = 0: $5.745\,T_{AC} + 6.495\,T_{AB} - 43\,300 = 0$ (1)

Σ(**j** components) = 0: $-4.817\,T_{AC} + 3.750\,T_{AB} = 0$ (2)

Solving for T_{AC} and T_{AB}: From (2): $T_{AB} = 1.285\,T_{AC}$ ♦

Substituting for T_{AB} into (1): $5.745\,T_{AC} + 6.495(1.285\,T_{AC}) = 43300$

$T_{AC} = 3074$ lb, $T_{AB} = 3949$ lb

4. (a) $\Sigma F_x = 0$: $O_x - 0.6428(3074) + 0.50(3949) = 0$

$O_X = 1.467$ lb

(b) $\Sigma F_y = 0$: $O_y - 0.7660(3074) - 0.8660(3949) = 0$

$O_y = 5775$ lb

(c) $\Sigma F_z = 0$: $O_Z - 5000 = 0$

$O_Z = 5000$ lb

(d) $\mathbf{R}_O = 1.47\vec{i} + 5780\vec{j} + 5000\vec{k}$ lb ♦

S 24.1

2.

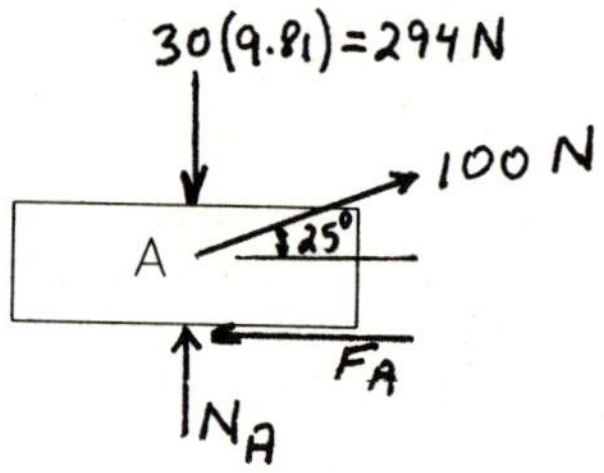

3. (a) $\Sigma F_y = 0$: $+\uparrow$ $N_A - 294 + 100 \sin 25° = 0$

$N_A = 251.7$ N

(b) $\Sigma F_x = 0$: $\xrightarrow{+}$ $100 \cos 25° - F_A = 0$

$F_A = 90.63$ N

4. $F_{max} = \mu_s N_A = 0.4N_A = 0.4(251.7) = 100.7$ N

Is $F_A \leq F_{max}$? yes $F_A = 90.6$ N The assumption of equilibrium was correct. ♦

S 24.2

1.

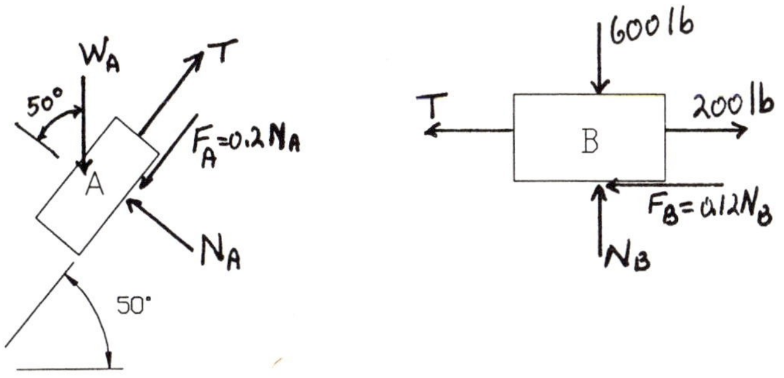

2. The unknowns on the FBDs: $W_A,\ N_A,\ T,\ N_B = 4$

   The number of available equilibrium equations = 4

3. Writing and solving the equilibrium equations for the weight of A: ◆

   Block B: $+\uparrow \Sigma F_y = 0: \quad N_B = 600 \text{ lb}$

   $\overset{+}{\rightarrow} \Sigma F_x = 0: \quad 200 - 0.12(600) - T = 0$

   $T = 128 \text{ lb}$

   Block A: $\overset{+}{\rightarrow} \Sigma F_x = 0: \quad 128\cos 50° - 0.2N_A\cos 50° - N_A\sin 50° = 0$

   $N_A = 91.97 \text{ lb}$

   $y'\ {}^{+}\!\nwarrow 50°\quad \Sigma F_y' = 0: \quad 91.97 - W_A\cos 50° = 0$

   $W_A = 143 \text{ lb}$

S 24.3

Method I

2.

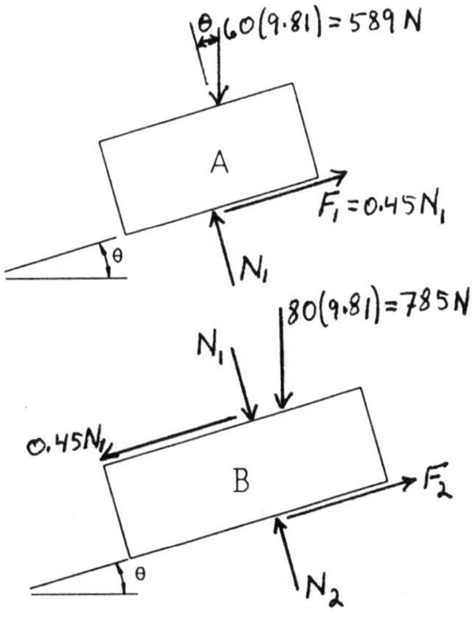

3. Writing and solving the equilibrium equations for θ:

   Block A: $y'\,+\!\nwarrow \quad \Sigma F_y' = 0: \quad N_1 - 589\cos\theta = 0$

   $\cos\theta = 0.0017N_1$

   $\overset{x'}{+\nearrow} \quad \Sigma F_x' = 0: \quad 0.45N_1 - 589\sin\theta = 0$

   $\sin\theta = 0.0008N_1$

   $$\tan\theta = \frac{\sin\theta}{\cos\theta} = \frac{0.0008N_1}{0.0017N_1} = 0.4494$$

   $$\theta = 24.2°$$

5.

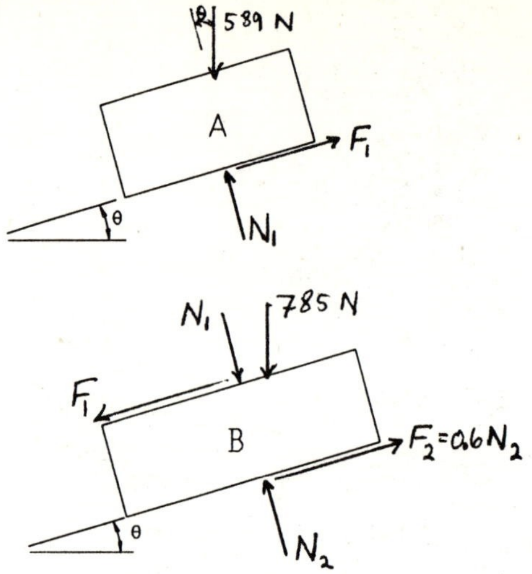

6. Writing and solving the equilibrium equations for θ:

Block A: $\Sigma F_{y'} = 0: \quad N_1 = 589\cos\theta$

$\Sigma F_{x'} = 0: \quad F_1 = 589\sin\theta$

Block B: $\Sigma F_{y'} = 0: \quad N_2 - 589\cos\theta - 785\cos\theta = 0$

$N_2 = 1374\cos\theta$

$\Sigma F_{x'} = 0: \quad 0.6(1374\cos\theta) - 589\sin\theta - 785\sin\theta = 0$

$824.4\cos\theta = 1374\sin\theta$

$\tan\theta = \frac{\sin\theta}{\cos\theta} = \frac{824.4}{1374} = 0.60 \qquad \theta = 31.0°$

7. θ = 24.2°, the smaller angle ◆

Method II

1. Writing the equilibrium equations and solving for all the unknowns:

From Step 2 of Method I: $\theta = 24.2°$, $N_1 = 589\cos 24.2 = 537.2$ N

Block B: $\Sigma F_{y'} = 0: \quad N_2 - 537.2 - 785\cos 24.2 = 0$

$N_2 = 1253$ N

$\Sigma F_{x'} = 0: \quad F_2 - 0.45(537.2) - 785\sin 24.2 = 0$

$F_2 = 563.5$ N

2. $F_{max} = \mu_s N = 0.6N_2 = 0.6(1253) = 751.8$ N

Is $F_2 \le F_{max}$? yes

θ = 24.2° ◆

S 25.1

1.

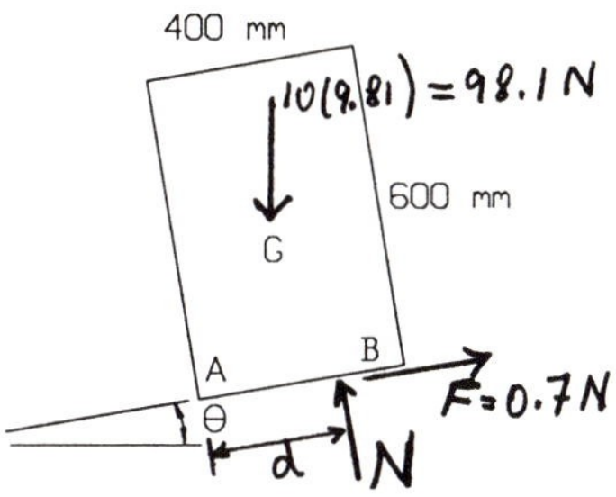

2. The equilibrium equations to solve for θ:

$\Sigma F_{y'} = 0: \quad N - 98.1\cos\theta = 0$

$N = 98.1\cos\theta$

$\Sigma F_{x'} = 0: \quad 0.7(98.1\cos\theta) - 98.1\sin\theta = 0$

$\frac{\sin\theta}{\cos\theta} = 0.70 = \tan\theta$

$\theta = 35.0^\circ$

3.

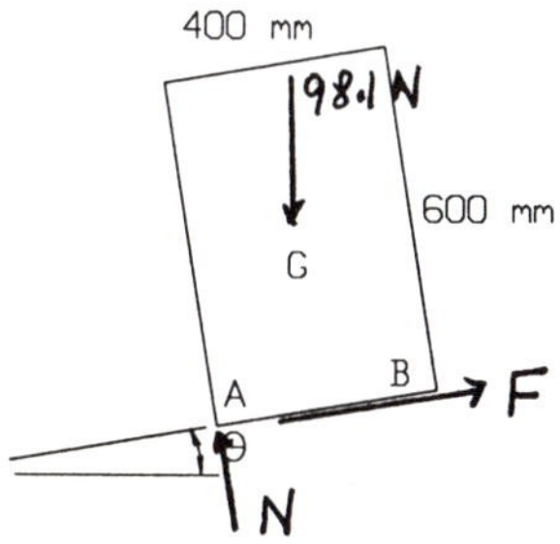

4. The equilibrium equations to solve for θ:

$\Sigma M_A = 0: \quad -98.1\cos\theta(200) + 98.1\sin\theta(300) = 0$

$300\sin\theta = 200\cos\theta$

$\frac{\sin\theta}{\cos\theta} = \frac{200}{300} = 0.667 = \tan\theta$

$\theta = 33.7^\circ$

5. Type of motion that impends: Tipping at $\theta = 33.7^\circ$ ♦

S 26.1

Part (a)

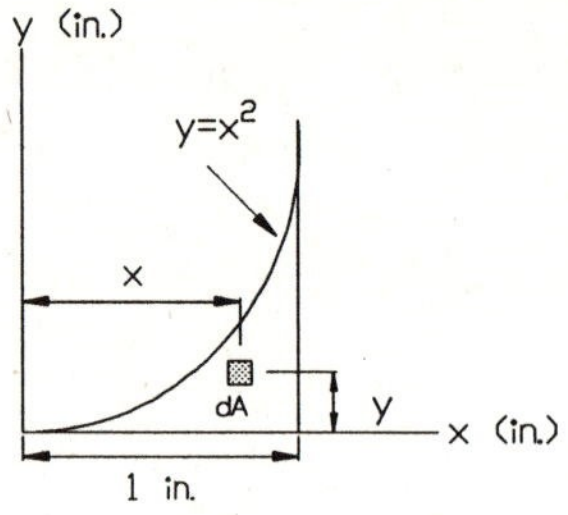

1. $dA = dy\,dx$

$$A = \int_0^1 \int_0^{x^2} dy\,dx = \int_0^1 \left(y\Big]_0^{x^2} \right) dx = \int_0^1 x^2 dx = \frac{x^3}{3}\Big]_0^1 = 0.333 \text{ in.}^2$$

2. $dQ_x = y\,dy\,dx$

$$Q_x = \int_0^1 \int_0^{x^2} y\,dy\,dx = \int_0^1 \left(\frac{y^2}{2}\Big]_0^{x^2} \right) dx = \int_0^1 \frac{x^4}{2} dx = \frac{x^5}{10}\Big]_0^1 = 0.100 \text{ in.}^3$$

3. $dQ_y = x\,dy\,dx$

$$Q_y = \int_0^1 \int_0^{x^2} x\,dy\,dx = \int_0^1 \left(xy\Big]_0^{x^2} \right) dx = \int_0^1 x^3 dx = \frac{x^4}{4}\Big]_0^1 = 0.250 \text{ in.}^3$$

4. $\bar{y} = Q_x / A = 0.100/0.333 = 0.300 \text{ in.}$ ♦

$\bar{x} = Q_y / A = 0.250/0.333 = 0.750 \text{ in.}$ ♦

Part (b)

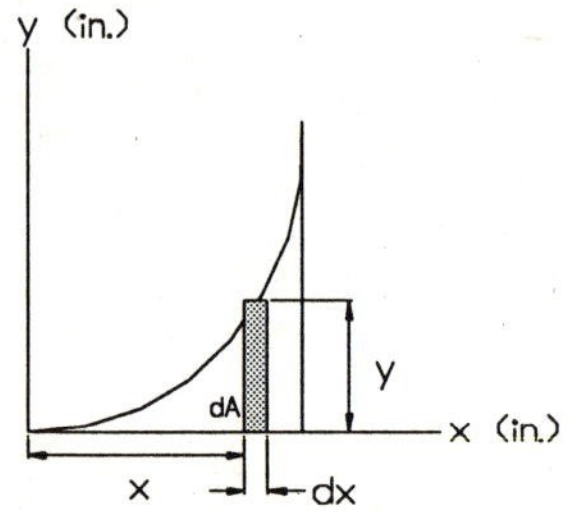

1. $dA = y\,dx = x^2 dx$

$$A = \int_0^1 x^2 dx = \frac{x^3}{3}\Big]_0^1 = 0.333 \text{ in.}^2$$

2. $dQ_x = y/2\ dA$

$$Q_x = \int_0^1 \frac{y}{2} dA = \int_0^1 \frac{x^4}{2} dx = \frac{x^5}{10}\Big]_0^1 = 0.100 \text{ in.}^3$$

3. $dQ_y = x\,dA$

$$Q_y = \int_0^1 x\,dA = \int_0^1 x^3\,dx = \frac{x^4}{4}\Big]_0^1 = 0.250\ \text{in.}^3$$

4. $\bar{y} = Q_x/A = 0.100/0.333 = 0.300$ in. ♦

$\bar{x} = Q_y/A = 0.250/0.333 = 0.750$ in. ♦

Part (c)

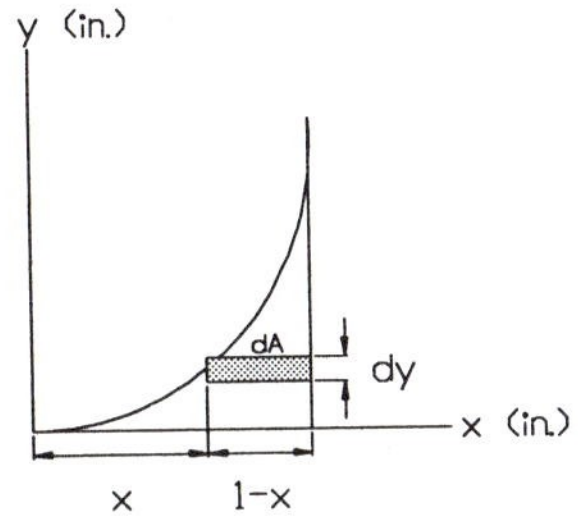

1. $dA = (1-x)\,dy = (1-\sqrt{y}\,)\,dy$

$$A = \int_0^1 (1-\sqrt{y})\,dy = y\Big]_0^1 - \frac{y^{1.5}}{1.5}\Big]_0^1 = 1.0 - 0.667 = 0.333\ \text{in.}^2$$

2. $dQ_x = y\,dA = y\,(1-\sqrt{y}\,)\,dy$

$$Q_x = \int_0^1 y(1-\sqrt{y})\,dy = \int_0^1 (y - y^{1.5})\,dy = \frac{y^2}{2}\Big]_0^1 - \frac{y^{2.5}}{2.5}\Big]_0^1 = 0.50 - 0.40 = 0.100\ \text{in.}^3$$

3. $dQ_y = x + \dfrac{1-x}{2}dA = \dfrac{1+x}{2}dA = \dfrac{1+\sqrt{y}}{2}(1-\sqrt{y})dy$

$$Q_y = \int_0^1 \left(\frac{1+\sqrt{y}}{2}\right)(1-\sqrt{y})\,dy = \int_0^1 \frac{1-y}{2}\,dy = \frac{y}{2}\Big]_0^1 - \frac{y^2}{4}\Big]_0^1 = 0.50 - 0.250 = 0.250\ \text{in.}^3$$

4. $\bar{y} = Q_x/A = 0.100/0.333 = 0.300$ in. ♦

$\bar{x} = Q_y/A = 0.250/0.333 = 0.750$ in. ♦

S 26.2

1.

Part	Area (mm^2)	$\bar{x}$ (mm)	$Q_y = A\bar{x}$ (mm^3)	$\bar{y}$ (mm)	$Q_x = A\bar{y}$ (mm^3)
A. Large triangle	$\frac{1}{2}(60)^2 = 1800$	-20.0	$1800(-20) = -36\,000$	20.0	$1800(20) = 36\,000$
B. Quarter circle	$\frac{\pi}{4}(20)^2 = 314$	$\frac{4(20)}{3\pi} = 8.49$	$314(8.49) = 2670$	$40 + \frac{4(20)}{3\pi} = 48.5$	$314(48.5) = 15230$
C. Small triangle	$\frac{1}{2}(20)(40) = 400$	$\frac{20}{3} = 6.67$	$400(6.67) = 2670$	$\frac{2}{3}(40) = 26.7$	$400(26.7) = 10680$
D. Quarter circle space	$-\frac{\pi}{4}(20)^2 = -314$	$\frac{-4(20)}{3\pi} = -8.49$	$-314(-8.49) = 2670$	$\frac{4(20)}{3\pi} = 8.49$	$-314(8.49) = -2670$
	$\Sigma A = 2200$		$\Sigma(A\bar{x}) = -28000$		$\Sigma(A\bar{y}) = 59200$

2. $\bar{x} = \Sigma Q_y / \Sigma A = \Sigma(A\bar{x})/\Sigma A = -28000/2200 = -12.7$ mm ◆

$\bar{y} = \Sigma Q_x / \Sigma A = \Sigma(A\bar{y})/\Sigma A = 59200/2200 = 26.9$ mm ◆

S 27.1

1.

Part	Vol. ($in.^3$)	$\bar{x}$ (in.)	$V\bar{x}$ ($in.^4$)	$\bar{y}$ (in.)	$V\bar{y}$ ($in.^4$)	$\bar{z}$ (in.)	$V\bar{z}$ ($in.^4$)
Block	$12 \times 12 \times 9 = 1296$	6.0	$1296(6) = 7776$	6.0	$1296(6) = 7776$	-4.5	$1296(-4.5) = -5832$
Cylindrical hole	$-\frac{\pi}{4}(4)^2(6) = -75.4$	9.0	$-75.4(9) = -679$	4.0	$-75.4(4) = -302$	-3.0	$-75.4(-3.0) = 226$
	$\Sigma V = 1221$		$\Sigma V\bar{x} = 7097$		$\Sigma V\bar{y} = 7474$		$\Sigma V\bar{z} = -5606$

2. $\bar{x} = \Sigma V\bar{x} / \Sigma V = 7097/1221 = 5.81$ in. ◆

$\bar{y} = \Sigma V\bar{y} / \Sigma V = 7474/1221 = 6.12$ in. ◆

$\bar{z} = \Sigma V\bar{z} / \Sigma V = -5606/1221 = -4.59$ in. ◆

S 28.1

1. For segment AB: $\bar{x} = 1/2\ \overline{AB} \sin 30° = \frac{1}{2}(40)(0.5) = 10.0$ mm

For segment BC: $\bar{x} = \overline{AB} \sin 30° + 1/2\ \overline{BC} \sin 10° = 40(0.5) + \frac{1}{2}(60)(0.1736)$
$= 20.0 + 5.209 = 25.21$ mm

2. For segment AB: $Q_X = L\bar{x} = 40(10.0) = 400 \text{ mm}^2$
For segment BC: $Q_X = L\bar{x} = 60(25.21) = 1512.6 \text{ mm}^2$

3. For segment AB: $A = 2\pi Q_X = 2\pi(400) = 2513 \text{ mm}^2$
For segment BC: $A = 2\pi Q_X = 2\pi(1512.6) = 9504 \text{ mm}^2$
Total Area = $2513 + 9503 = 12.02 \times 10^3 \text{ mm}^2$ ♦

S 29.1

1. Wood block: Weight = $\gamma V = 40(1.0 \times 1.0 \times 0.75) = 30.0$ lb
Wood cylinder: Weight = $\gamma V = 40\left[\frac{\pi}{4}(0.333)^2 \times 0.5\right] = 1.742$ lb
Metal cylinder: Weight = $\gamma V = 1146\left[\frac{\pi}{4}(0.333)^2 \times 0.5\right] = 50.0$ lb

2.

Part	Weight (lb)	$\bar{x}$ (in.)	$W\bar{x}$ (lb•in.)	$\bar{y}$ (in.)	$W\bar{y}$ (lb•in.)	$\bar{z}$ (in.)	$W\bar{z}$ (lb•in.)
Wood Block	30.0	6.0	30.0(6.0) = 180	6.0	30.0(6.0) = 180	4.5	30.0(4.5) = 135
Wood cylinder	−1.742	9.0	−1.742(9.0) = −15.68	4.0	−1.742(4.0) = −6.968	6.0	−1.742(6.0) = −10.45
Metal cylinder	50.0	9.0	50.0(9.0) = 450	4.0	50.0(4.0) = 200	6.0	50.0(6.0) = 300
	$\Sigma W = 78.26$		$\Sigma W\bar{x} = 614.3$		$\Sigma W\bar{y} = 373$		$\Sigma W\bar{z} = 424.5$

3. $\bar{x} = \Sigma W\bar{x} / \Sigma W = 614.3/78.26 = 7.85$ in. ♦
$\bar{y} = \Sigma W\bar{y} / \Sigma W = 373/78.26 = 4.77$ in. ♦
$\bar{z} = \Sigma W\bar{z} / \Sigma W = 424.5/78.26 = 5.42$ in. ♦

S 30.1

1. $\gamma = \rho g = (1000 \text{ kg/m}^3)(9.81 \text{ m/s}^2) = 9810 \text{ N/m}^3$
2. p_{bottom} = water depth x $\gamma = 1.2(9810) = 11770 \text{ N/m}^2$
3.

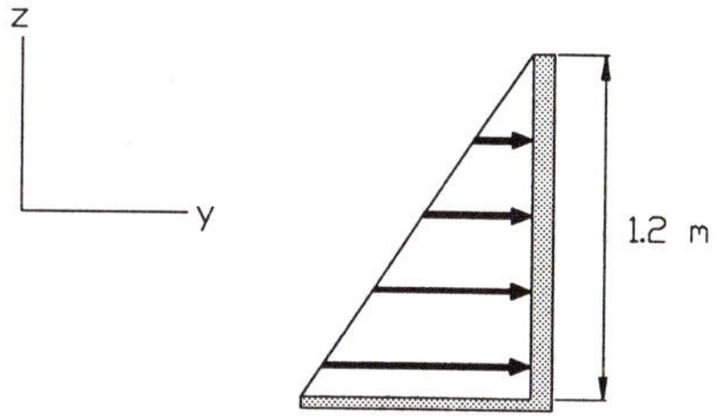

4. Resultant Force = $\frac{1}{2}(11770)(0.5)(1.2) = 3531$ N ♦
5. $z = \frac{1}{3}(1.2) = 0.40$ m ♦

S 31.1

Part (a)

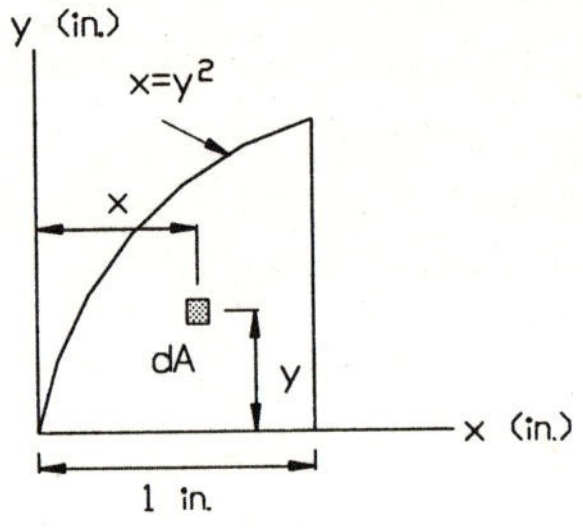

1. dA = dy dx

$$I_x = \int_0^1 \int_0^{\sqrt{x}} y^2 dy\, dx = \int_0^1 \frac{y^3}{3}\Big]_0^{\sqrt{x}} dx = \int_0^1 \frac{x^{1.5}}{3} dx = \frac{x^{2.5}}{7.5}\Big]_0^1$$ ♦

$$= 0.133 \text{ in.}^4$$

Part (b)

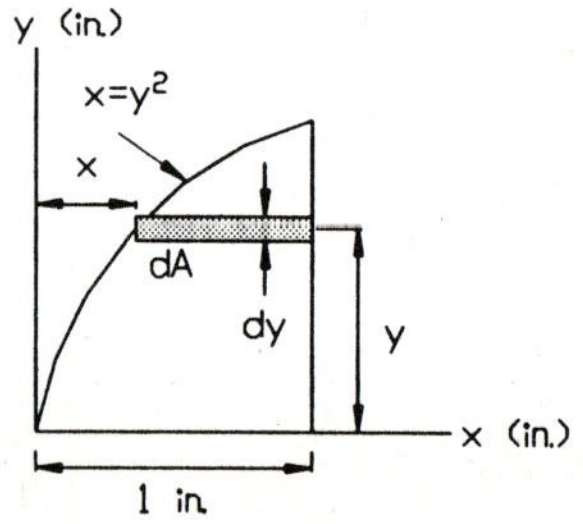

1. $dA = (1-x)\,dy = (1-y^2)\,dy$

$$I_x = \int_0^1 y^2(1-y^2)dy = \int_0^1 (y^2 - y^4)dy = \frac{y^3}{3}\Big]_0^1 - \frac{y^5}{5}\Big]_0^1 = \frac{1}{3} - \frac{1}{5}$$

$$= 0.133 \text{ in.}^4$$

Part (c)

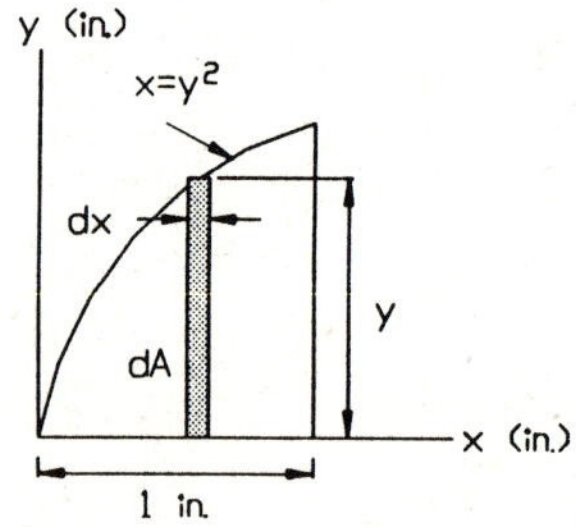

1. $dI_x = dx(y^3/3) = dx\dfrac{x^{3/2}}{3}$

$$I_x = \int_0^1 \frac{x^{3/2}}{3} dx = \int_0^1 \frac{x^{1.5}}{3} dx = \frac{x^{2.5}}{7.5}\Big]_0^1 = 0.133 \text{ in.}^4$$

Part (d)

1. (a) $A = \int_0^1 \int_0^{\sqrt{x}} dy\,dx = \int_0^1 y \Big]_0^{\sqrt{x}} dx = \int_0^1 x^{0.5}\,dx = \frac{x^{1.5}}{1.5}\Big]_0^1 = 0.667 \text{ in.}^2$

   (b) $k_x = \sqrt{\frac{I_x}{A}} = \sqrt{\frac{0.133}{0.667}} = 0.447 \text{ in.}$ ◆

S 31.2

1. (a) $A = 1/2(bh) = \frac{1}{2}(40)(30) = 600 \text{ mm}^2$

   (b) $I_x = \bar{I}_x + Ad^2 = \frac{bh^3}{36} + Ad^2 = \frac{40(30)^3}{36} + 600(20)^2 = 30\,000 + 240\,000 = 270 \times 10^3 \text{ mm}^4$

   (c) $I_y = \bar{I}_y + Ad^2 = \frac{hb^3}{36} + Ad^2 = \frac{30(40)^3}{36} + 600\left(20 + \frac{40}{3}\right)^2 = 53\,333 + 666\,667 = 720 \times 10^3 \text{ mm}^4$

2. (a) $A = \pi R^2/2 = \pi(20)^2/2 = 628.3 \text{ mm}^2$

   (b) $I_x = \bar{I}_x + Ad^2 = 0.1098R^4 + Ad^2 = 0.1098(20)^4 + 628.3\left[30 + \frac{4(20)}{3\pi}\right]^2$
   $= 17\,568 + 930\,730 = 948.3 \times 10^3 \text{ mm}^4$

   (c) $I_y = \bar{I}_y + Ad^2 = \frac{\pi R^4}{8} + Ad^2 = \pi(20)^4/8 + 628.3\left(20 + \frac{40}{2}\right)^2$
   $= 62\,832 + 1\,005\,280 = 1.068 \times 10^6 \text{ mm}^4$

3. $I_x = \Sigma I_x = 270 \times 10^3 + 948.3 \times 10^3 = 1.218 \times 10^6 \text{ mm}^4$ ◆

   $I_y = \Sigma I_y = 720 \times 10^3 + 1.068 \times 10^6 = 1.788 \times 10^6 \text{ mm}^4$ ◆

S 32.1

1. $I_{xy} = \bar{I}_{xy} + A\bar{x}\bar{y} = 0 + 8(6)(6+4)(3) = 1440 \text{ in.}^4$

2. $I_{xy} = \bar{I}_{xy} + A\bar{x}\bar{y} = \frac{-b^2h^2}{72} + A\bar{x}\bar{y} = \frac{-8^2(4^2)}{72} + \frac{1}{2}(8)(4)\left[6 + \frac{2}{3}(8)\right]\left[6 + \frac{1}{3}(4)\right]$
   $= -14.22 + 16(11.34)(7.333) = 1316 \text{ in.}^4$

3. $I_{xy} = \Sigma I_{xy} = 1440 + 1316 = 2760 \text{ in.}^4$ ◆

S 33.1

1.

$$\bar{I}_x = \frac{bh^3}{36} = \frac{4(3)^3}{36} = 3.0 \text{ in.}^4$$

$$\bar{I}_y = \frac{hb^3}{36} = \frac{3(4)^3}{36} = 5.33 \text{ in.}^4$$

$$\bar{I}_{xy} = -\frac{b^2h^2}{72} = -\frac{4^2(3)^2}{72} = -2.0 \text{ in.}^4$$

2.

$$\bar{I}_1 = \frac{\bar{I}_x + \bar{I}_y}{2} + \sqrt{\left(\frac{\bar{I}_x - \bar{I}_y}{2}\right)^2 + \bar{I}_{xy}^2} = \frac{3.0 + 5.33}{2} + \sqrt{\left(\frac{3.0 - 5.33}{2}\right)^2 + (-2.0)^2}$$ ♦

$$= 4.165 + \sqrt{5.357} = 6.48 \text{ in.}^4$$

$$\bar{I}_2 = \frac{\bar{I}_x + \bar{I}_y}{2} - \sqrt{\left(\frac{\bar{I}_x - \bar{I}_y}{2}\right)^2 + \bar{I}_{xy}^2} = 4.165 - \sqrt{5.357} = 1.85 \text{ in.}^4$$ ♦

3.

$$\tan 2\theta = -\frac{2\bar{I}_{xy}}{\bar{I}_x - \bar{I}_y} = -\frac{2(-2)}{3.0 - 5.33} = -1.717$$

$$2\theta = -59.8^\circ$$

$$\theta_1 = -29.9^\circ$$ ♦

$$\theta_2 = \theta_1 + 90^\circ = -29.9^\circ + 90^\circ = 60.1^\circ$$ ♦